Prepared in cooperation with the Northern Cheyenne Tribe

Estimates of the Volume of Water in Five Coal Aquifers, Northern Cheyenne Indian Reservation, Southeastern Montana

Scientific Investigations Report 2012–5209

U.S. Department of the Interior
U.S. Geological Survey

Estimates of the Volume of Water in Five Coal Aquifers, Northern Cheyenne Indian Reservation, Southeastern Montana

By L.K. Tuck, Daniel K. Pearson, M.R. Cannon, and DeAnn M. Dutton

Prepared in cooperation with the Northern Cheyenne Tribe

Scientific Investigations Report 2012–5209

U.S. Department of the Interior
U.S. Geological Survey

U.S. Department of the Interior
SALLY JEWELL, Secretary

U.S. Geological Survey
Suzette M. Kimball, Acting Director

U.S. Geological Survey, Reston, Virginia: 2013

For more information on the USGS—the Federal source for science about the Earth, its natural and living resources, natural hazards, and the environment, visit http://www.usgs.gov or call 1–888–ASK–USGS.

For an overview of USGS information products, including maps, imagery, and publications,
visit http://www.usgs.gov/pubprod

To order this and other USGS information products, visit http://store.usgs.gov

Suggested citation:
Tuck, L.K., Pearson, D.K., Cannon, M.R., and Dutton, D.M., 2013, Estimates of the volume of water in five coal aquifers, Northern Cheyenne Indian Reservation, southeastern Montana: U.S. Geological Survey Scientific Investigations Report 2012–5209, 26 p., http://pubs.usgs.gov/sir/2012/5209/.

Acknowledgments

The Northern Cheyenne Tribe is gratefully acknowledged for field assistance and information about the study area. We also gratefully acknowledge the invaluable contribution of Edward L. Heffern, Bureau of Land Management (retired), who provided personal observations about the geology and coal stratigraphy of the study area.

Contents

Figures

Tables

Conversion Factors, Datum, Transmissivity, Water-Year Definition, and Abbreviations

Multiply	By	To obtain
acre	4,047	square meter (m^2)
acre	0.004047	square kilometer (km^2)
acre-foot (acre-ft)	1,233	cubic meter (m^3)
cubic foot	0.02832	cubic meter (m^3)
cubic foot per second (ft^3/s)	0.02832	cubic meter per second (m^3/s)
inch (in.)	2.54	centimeter (cm)
inch (in.)	25.4	millimeter (mm)
foot (ft)	0.3048	meter (m)
foot per day (ft/d)	0.3048	meter per day (m/d)
foot squared per day (ft^2/d)	0.09290	meter squared per day (m^2/d)
gallon per minute (gal/min)	0.06309	liter per second (L/s)
mile (mi)	1.609	kilometer (km)
square mile (mi^2)	2.590	square kilometer (km^2)

Temperature in degrees Celsius (°C) may be converted to degrees Fahrenheit (°F) as follows: °F=(1.8×°C)+32.

Vertical coordinate information is referenced to the North American Vertical Datum of 1988 (NAVD 88) and the National Geodetic Vertical Datum of 1929 (NGVD 29). Horizontal coordinate information is referenced to the North American Datum of 1983 (NAD 83) and the North American Datum of 1927 (NAD 27).

Altitude, as used in this report, refers to distance above the vertical datum.

Transmissivity: The standard unit for transmissivity is cubic foot per day per square foot times foot of aquifer thickness [(ft^3/d)/ft^2]ft. In this report, the mathematically reduced form, foot squared per day (ft^2/d), is used for convenience.

Water-year definition: Water year is the 12-month period from October 1 through September 30 of the following calendar year. The water year is designated by the calendar year in which it ends. For example, water year 2010 is the period from October 1, 2009, through September 30, 2010.

Abbreviations Used in This Report:

BLM — Bureau of Land Management

CBM — coal-bed methane

CD–ROM — compact disc—read-only memory

DEM — digital-elevation model

ESRI — Environmental Systems Research Institute

GIS — geographic information system

GWIC — Ground Water Information Center, Montana Bureau of Mines and Geology (database)

IDW — inverse distance weighting

MBMG — Montana Bureau of Mines and Geology

NWIS — National Water Information System, Groundwater Information and Data (database), U.S. Geological Survey

USGS — U.S. Geological Survey

Estimates of the Volume of Water in Five Coal Aquifers, Northern Cheyenne Indian Reservation, Southeastern Montana

By L.K. Tuck, Daniel K. Pearson, M.R. Cannon,[1] and DeAnn M. Dutton

Abstract

Water is one of the Northern Cheyenne Tribe's most valuable natural resources—vital to the health and economic welfare of the Northern Cheyenne people. The Tongue River Member of the Tertiary Fort Union Formation is the primary source of groundwater in the Northern Cheyenne Indian Reservation in southeastern Montana. Coal beds within this formation generally contain the most laterally extensive aquifers in much of the reservation. The U.S. Geological Survey, in cooperation with the Northern Cheyenne Tribe, conducted a study to estimate the volume of water in five coal aquifers.

This report presents estimates of the volume of water in five coal aquifers in the eastern and southern parts of the Northern Cheyenne Indian Reservation. The Canyon, Wall, Pawnee, Knobloch, and Flowers-Goodale coal beds in the Tongue River Member of the Tertiary Fort Union Formation were investigated. Only conservative estimates of the volume of water in the Canyon, Wall, Pawnee, Knobloch, and Flowers-Goodale coal aquifers are presented because (1) the subsurface extent of the coal beds are not well defined, (2) in some instances, well and drill-hole data were widely spaced and not well distributed, (3) of the possibility that some coal beds split, merge, or pinch out laterally, and (4) water-level data for the five aquifers were scarce.

The volume of water in the Canyon coal was estimated to range from about 10,400 acre-feet (75 percent saturated) to 3,450 acre-feet (25 percent saturated). The smaller estimates of water in the Canyon coal (75 to 25 percent saturation) are considered more reasonable; 100 percent saturation was not considered probable within the study area. However, estimates of the volume of water in the Canyon coal might have large errors and need to be used with caution because the water-level data needed to define the volume of water were unavailable.

The volume of water in the Wall coal was estimated to range from about 14,200 acre-feet (100 percent saturated) to 3,560 acre-feet (25 percent saturated). Water-level data indicate that the Wall coal was both unconfined and confined within and near the study area. Thus, the estimates of the volume of water in the Wall coal probably are reasonable.

The volume of water in the Pawnee coal was estimated to range from about 9,440 acre-feet (100 percent saturated) to 2,360 acre-feet (25 percent saturated). Water-level data from one well and information from one geologist's log indicated that the Pawnee coal probably was fully saturated and might have been under confined conditions in the study area. However, estimates of the volume of water in the Pawnee coal might have large errors and need to be used with caution because the water-level data needed to define the volume of water were largely unavailable.

The volume of water in the Knobloch coal was estimated to range from about 38,700 acre-feet (100 percent saturated) to 9,680 acre-feet (25 percent saturated). Water-level data indicate that the Knobloch coal was both unconfined and confined within and near the study area. Thus, the estimates of the volume of water in the Knobloch coal probably are reasonable.

The volume of water in the Flowers-Goodale coal was estimated to be about 35,800 acre-feet (100 percent saturated). Water-level data and information from one geologist's log indicate that the Flowers-Goodale coal was confined at these wells. Also, because this coal generally is deeply buried in the study area, the Flowers-Goodale coal was assumed to be confined throughout its extent in the study area. Thus, the estimates of the volume of water in the Flowers-Goodale coal probably are reasonable.

Sufficient data are needed to accurately characterize coal-bed horizontal and vertical variability, which is highly complex both locally and regionally. Where data points are widely spaced, the reliability of estimates of the volume of coal beds is decreased. Additionally, reliable estimates of the volume of water in coal aquifers depend heavily on data about water levels and data about coal-aquifer characteristics. Because the data needed to define the volume of water were sparse, only conservative estimates of the volume of water in the five coal aquifers are presented in this report. These estimates need to be used with caution and mindfulness of the uncertainty associated with these estimates.

---

[1]U.S. Geological Survey, retired.

Introduction

Water is one of the Northern Cheyenne Tribe's most valuable natural resources—vital to the health and economic welfare of the Northern Cheyenne people. Except for the Tongue River and Rosebud Creek (fig. 1), surface water in many parts of the Northern Cheyenne Indian Reservation is not available for use. Although small ephemeral streams can provide water for livestock and some agriculture, most water for domestic, livestock, and municipal use is obtained from wells (Matson and Blumer, 1973; Barbara A. Burkland, U.S. Environmental Protection Agency, oral commun., 2010). The Tongue River Member of the Tertiary Fort Union Formation is the primary source of groundwater in the Northern Cheyenne Indian Reservation, in southeastern Montana. Coal beds within this formation generally contain the most laterally extensive aquifers; coal aquifers are readily used and can provide abundant water to wells and springs (Woessner and others, 1981). In much of the reservation, a practical alternative to groundwater from the Tongue River Member does not exist.

Coal-bed methane (CBM) production is an important industry in southeastern Montana. Potential reserves of CBM might exist in the southern and eastern parts of the Northern Cheyenne Indian Reservation and in adjacent areas where some of the same coal aquifers in the Tongue River Member extend beyond the boundaries of the reservation (Biewick and McLellan, 1990). To extract this methane, groundwater is pumped from the coal aquifer so that water levels in wells (hydraulic head) are reduced sufficiently to release methane stored in the coal aquifer; the methane is in solution in the water (Wo and others, 2004). The extraction and subsequent management of CBM-produced water has raised concerns about the potential reduction of groundwater supplies caused by lowering of water levels and the potential effects of the disposal of produced water on surface water and soils. After 10 years of CBM production in coal fields near Decker, Mont. (fig. 1), Meredith and others (2010) reported that water levels in some of the coal aquifers in the production field decreased 150 to 600 feet (ft), and that as far as 1.0 to 1.5 miles (mi) beyond the boundaries of the production field, water levels had decreased by 20 ft. Likewise, flow from springs and water available in wells might be diminished proportionally to the decrease in hydraulic head in the aquifer (Wheaton and Donato, 2004). Where pumping has lowered water levels below the top of a confined coal aquifer (below the base of a confining or leaky-confining layer), the coal aquifer then becomes unconfined and dewatering of the aquifer occurs. Dewatering reduces the saturated thickness of an aquifer and can affect the productivity of the aquifer (Slagle and others, 1985).

Additionally, in 2010, the State of Montana approved the lease of coal resources of about 8,300 acres that are approximately 3 mi east of the Tongue River (Montana Department of Natural Resources, 2002). The Tongue River is the eastern boundary of the Northern Cheyenne Indian Reservation. Some of the coal beds targeted for that development under this lease might be contiguous with those in the eastern part of the reservation (Heffern and others, 1993).

Information about the volume of water in coal aquifers in the reservation was needed because possible development of coal and CBM could deplete the Tribe's groundwater resources by lowering water levels. This potential depletion also raised concerns about groundwater rights and that loss of water from coal aquifers would affect traditional land and water uses. To enable the Northern Cheyenne Tribe to manage its groundwater resources, the amount of groundwater in the coal aquifers underlying the reservation needed to be estimated. Consequently, the U.S. Geological Survey (USGS), in cooperation with the Northern Cheyenne Tribe, conducted a study to estimate the volume of water in five coal aquifers. The eastern and southern parts of the reservation were selected for study because some coal aquifers in this area might be more likely to be affected by CBM development.

Purpose and Scope

This report presents estimates of the volume of water in five coal aquifers in the eastern and southern parts of the Northern Cheyenne Indian Reservation. The Canyon, Wall, Pawnee, Knobloch, and Flowers-Goodale coal beds in the Tongue River Member of the Tertiary Fort Union Formation were investigated (stratigraphic nomenclature from Flores and others, 2010). These five coal beds are known to yield water in the study area; thus, coal aquifers contained in these coal beds are the focus of this report. Other coal beds within the Tongue River Member in the study area also probably yield water in usable quantities. However, these other coal beds were not investigated and their water-yielding properties are unknown.

To define the volume of water in each coal aquifer, existing data from lithologic logs from various sources, along with geologic maps of the study area, were compiled. Extent and thickness data for each coal bed were then used to estimate the volume of coal for each of the five coal beds by using computer interpolation methods. The volume of water in each of the five coal aquifers was estimated by assuming that each coal bed was fully saturated (thickness of the coal bed was equal to the thickness of the coal aquifer, except for the Canyon coal). Estimates of the volume of water in these coal aquifers were determined from existing data about hydraulic properties of coal beds in areas of Montana, North Dakota, and Wyoming. Existing water-level data measured from wells in the study area and from nearby areas were used to determine if coal beds were unconfined or confined and also to evaluate the percentage of saturation for each coal bed and, thus, to estimate the volume of water in each coal aquifer.

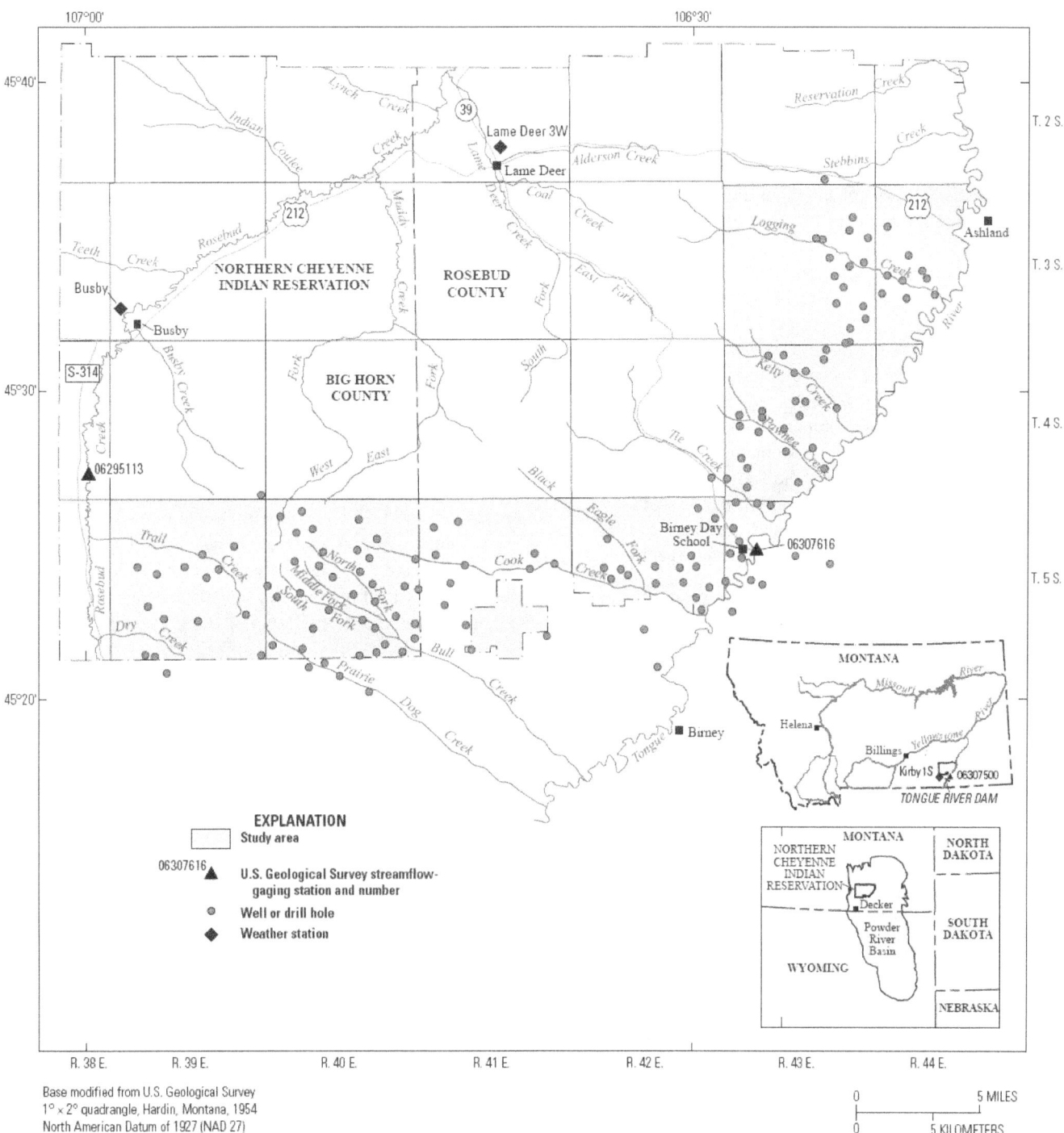

Figure 1.   Location of the study area, southeastern Montana.

Description of Study Area

The Northern Cheyenne Indian Reservation (fig. 1) encompasses about 445,000 acres in Big Horn and Rosebud Counties in southeastern Montana (Wo and others, 2004). The reservation lies on an unglaciated, semiarid, rolling plain that is underlain by the Tertiary Fort Union Formation and is dissected by many small ephemeral or intermittent streams. Grass-covered rangeland used to raise cattle is interspersed with farmland (Woods and others, 2002); hay is raised as the principal crop along the valley bottoms of the Tongue River and Rosebud and Lame Deer Creeks. Native grasslands are extensive, especially in areas of steep or rugged terrain. Pine and juniper forests predominate in upland areas; grazing in these areas can be limited because of rough terrain and lack of water (Woods and others, 2002).

The Tongue River is an important source of surface water in this part of southeastern Montana. Streamflow in the Tongue River is regulated by the Tongue River Dam (fig. 1), about 73 river miles upstream from Ashland (Montana Department of Natural Resources and Conservation, 1976). For water years 1939–2010, the annual mean streamflow at Tongue River at Tongue River Dam, near Decker, Mont. (streamflow-gaging station 06307500), was 430 cubic feet per second (ft^3/s). For water years 1980–2010 the mean annual streamflow at Tongue River at Birney Day School Bridge, near Birney, Mont. (streamflow-gaging station 06307616), was 380 ft^3/s (U.S. Geological Survey, 2010a).

Rosebud Creek, also an important source of surface water, flows across the western and northern parts of the reservation. For water years 1980–2010, the mean annual streamflow of Rosebud Creek at reservation boundary, near Kirby, Mont. (streamflow-gaging station 06295113), was 6.29 ft^3/s (U.S. Geological Survey, 2010a).

The climate of the study area is characterized by cold, dry winters and hot, moderately dry summers. December and January typically are the coldest months, whereas July and August typically are the warmest months. At Busby, Kirby 1S, and Lame Deer 3W weather stations (fig. 1), the mean monthly minimum temperature in December and January ranged from about 4 to 10°F and the mean monthly maximum temperature in July and August ranged from about 86 to 88°F. Mean annual precipitation at these three stations ranged from about 14 to 19 inches (in.); about 45 to 55 percent of the mean precipitation falls in April, May, June, and July (various periods of record from 1907 to 2010; Western Regional Climate Center, 2010).

Geohydrologic Framework

Geologic Setting

The Northern Cheyenne Indian Reservation is located in the northwestern part of the Powder River Basin (fig. 1), which is an asymmetrical structural basin. Within Montana, the axis of the Powder River Basin trends northeast-southwest, approximately along the valley of the Tongue River. Along the northwestern margin of the basin, rocks of the Fort Union Formation dip less than 1 degree to the southeast. The Powder River Basin covers more than 21,600 square miles (mi^2) in Montana and Wyoming (Heffern and others, 2007). A shallow syncline plunges to the south through the center of the reservation (Wo and others, 2004), and a system of northeast-trending normal faults in the northeastern part of the reservation formed as a result of strike-slip movements along the boundaries of basement rocks (Culberson and Saperstone, 1987a, b; Wo and others, 2004). In places, faults can offset strata by as much as 60 to 160 ft (Heffern, 1980; Woessner and others, 1981).

The Fort Union Formation of Tertiary age consists of three members—the Tullock, Lebo Shale, and Tongue River Members; the Lebo Shale and Tullock Members underlie the Tongue River Member. Generally, the Tullock Member consists mostly of thin-bedded siltstone and sandstone and only local stringers of coal (Hansen and Culbertson, 1985; Culbertson, 1987). The Lebo Shale member consists mostly of dark shale and mudstone and contains a few thin beds of sandstone and coal, whereas the Tongue River Member consists of interbedded sandstone, siltstone, shale, coal, and a few lenses of limestone. The Fort Union Formation can be as much as 3,900 ft thick in the Powder River Basin (Lewis and Roberts, 1978) but ranges from about 800 to 1,700 ft thick in the reservation (Hopkins, 1973; Vuke and others, 2001a, b).

Sedimentary rocks of the Tongue River Member (fig. 2) are exposed at the surface in most of the study area and underlie the entire study area. These rocks were deposited in the Powder River Basin, which formed as a result of intermittent crustal movements throughout Late Cretaceous and Tertiary time. The depositional environments were mainly fluvial systems consisting of braided and meandering streams in the center of the basin and alluvial fans along the western basin margin. Coals formed in numerous and extensive peat mires or swamps in fluvial flood plains, abandoned fluvial channels, and interchannel environments (Flores and Ethridge, 1985; Flores, 1986; Heffern and others, 2007; Flores and others, 2010).

In the Birney-Decker area (fig. 1), the Tongue River Member can locally contain as many as 20 coal beds. However, the beds split, merge, or pinch out laterally within relatively short distances (about 3 to 5 mi), which complicates correlation throughout this part of Montana (Culbertson and Saperstone, 1987a, b; Culbertson, 1987). The interval of the Tongue River Member present on the reservation contains about 15 informally named coal beds (Woessner and others, 1981; Culbertson, 1987; Wo and others, 2004) and several local unnamed coal beds. Some of the coal beds are quite laterally extensive. For example, east of the reservation, the Knobloch coal, which represents a long period of continued peat deposition, underlies an area of 270 mi^2 (Culbertson and Saperstone, 1987b; McLellan and others, 1990).

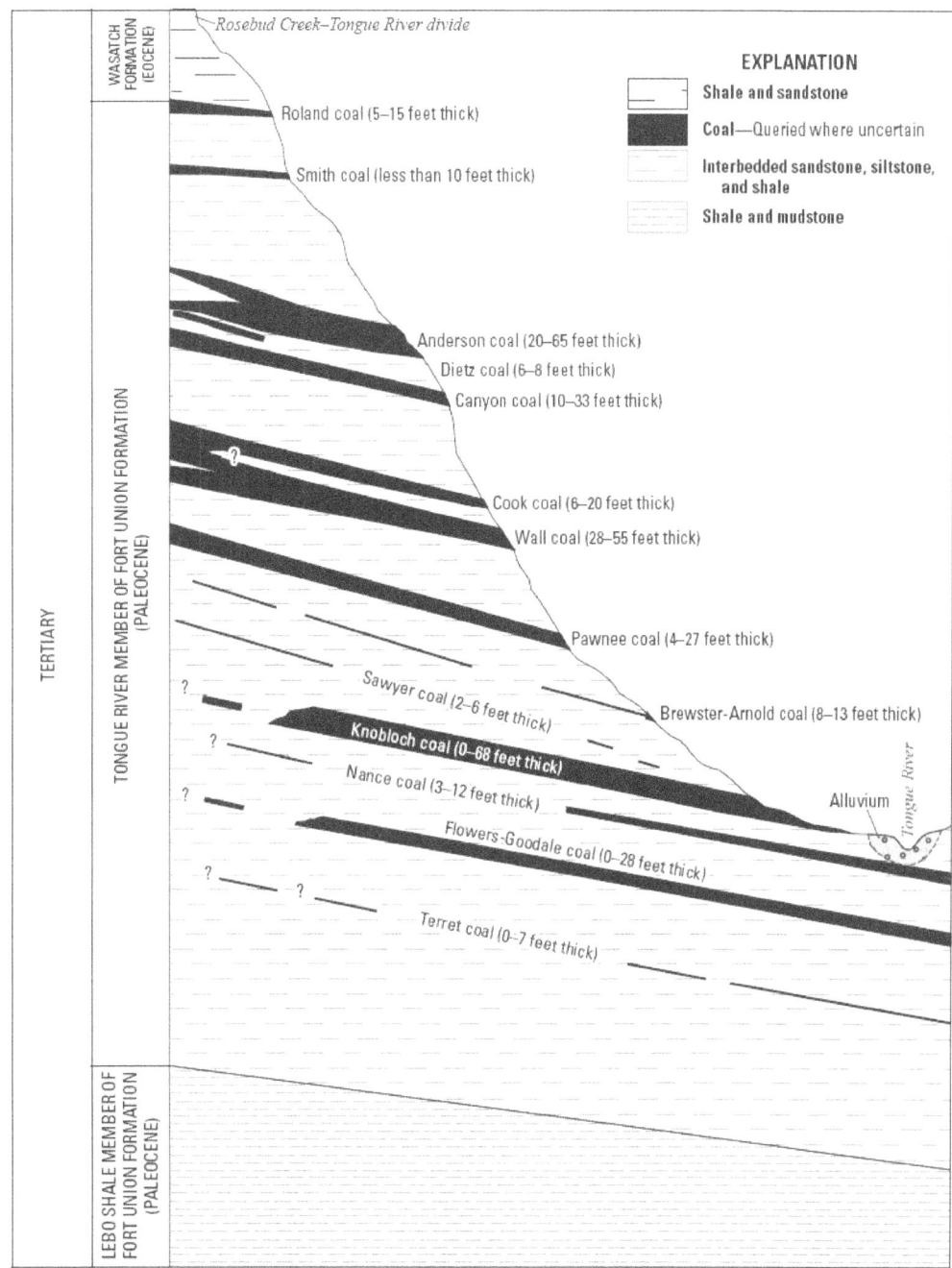

Figure 2. Idealized stratigraphic section of the Tongue River Member of the Fort Union Formation from the Rosebud Creek–Tongue River divide eastward to the Tongue River (thickness and extent information modified from Woessner and others, 1981; Culbertson, 1987).

Along the valleys of Rosebud Creek and the Tongue River, much of the Tongue River Member has been removed by erosion. Because the reservation is highly dissected by ephemeral and perennial streams, many coal beds crop out only as erosional remnants along some of the ridges and higher plateaus.

Overlying the Tongue River Member, along the drainage divide between Rosebud Creek and the Tongue River, Tertiary-aged sedimentary rocks of the Wasatch Formation cap some of the higher buttes and ridges (fig. 2). The Wasatch Formation, which can be as much as 500 ft thick, consists of grayish-brown and gray shale, some carbonaceous shale, and thin beds of brown calcareous sandstone. The contact of the Fort Union and Wasatch Formations is placed at the top of the Roland coal (Hopkins, 1973; Matson and Blumer, 1973).

Thick beds of red clinker—formed by the burning, welding, and melting of sedimentary rock above and below coal beds—are present where coal beds have burned along their outcrops. Most coal beds have burned in place and, consequently, formed distinctive red, erosion-resistant clinker (Woessner and others, 1981; Heffern and others, 2007). Generally, clinker erodes more slowly than other rocks of the Tongue River Member and Wasatch Formation and, thus, tends to control topography by capping plateaus and hilltops to form escarpments (Heffern and Coates, 2004). About 7 percent of the Powder River Basin and about 30 percent of the reservation is covered by outcrops of clinker (Heffern, 1980; Heffern and others, 2007).

The valleys of Rosebud Creek, Tongue River, and some of their larger tributaries contain Quaternary alluvium derived from sandstone, shale, and clinker of the Fort Union and Wasatch Formations. Alluvium underlying these stream valleys consists of gravel, sand, silt, and clay; in some areas, the alluvium can be thick. These streams have eroded through parts of Tongue River Member and exposed coal along valleys and in coulees (Woessner and others, 1981; Wheaton and Donato, 2004).

Hydrologic Setting

Principal aquifers in the study area include alluvium, clinker, and sandstone and coal. Alluvium, which can be as much as 80 ft thick, is primarily located in the valleys of the Tongue River and Rosebud and Lame Deer Creeks (Hopkins, 1973); alluvium also extends along the channels of smaller intermittent streams. Locally, alluvium can contain sufficient saturated sand and gravel to be an important source of water for domestic and livestock use and limited irrigation. Wells completed in alluvium typically yield 10 to 20 gallons per minute (gal/min) (Hopkins, 1973; Woessner and others, 1981).

Clinker is highly fractured, porous, and permeable; these characteristics allow infiltration of precipitation and percolation of water and flow to springs, which typically can be found along the base of clinker outcrops. Lateral flow from adjacent coal beds also recharges clinker (Hopkins, 1973; Woessner and others, 1981; Cannon, 1982). Conversely, clinker can form discontinuous, generally perched aquifers because underlying and overlying sandstone and shale typically were altered (hardened and perhaps melted by baking) and now can be less permeable than the fractured clinker (Woessner and others, 1981). Few wells are completed in clinker because of the extreme difficulty of drilling and because of its small saturated thickness. In many areas, water in clinker can move through talus or colluvium into adjacent alluvium without reaching land surface to issue as a spring. Clinker can be a reliable source of water for domestic and livestock use (Woessner and others, 1981; Cannon, 1982; McClymonds, 1982).

Sandstone and coal aquifers of the Tongue River Member are present throughout the reservation and collectively compose the most laterally extensive hydrogeologic units used for domestic and livestock wells (Woessner and others, 1981). Sandstone beds generally are lenticular and interbedded with shale, which can result in discontinuous aquifers. However, the many lenticular sandstone beds, combined with the many coal beds, create hydrogeologic units where water supplies can be obtained (Woessner and others, 1981). Some of the sandstone beds can be as much as 100 ft thick with porosities as much as 30 percent (Wo and others, 2004). Yields of wells in sandstone and coal aquifers generally range from about 2 to 50 gal/min (Hopkins, 1973: Woessner and others, 1981; McClymonds, 1982). Conversely, some sandstone can be relatively impermeable (probably leaky-confining layers), and both the sandstone and coal can be dry depending on the location of the beds (Hopkins, 1973; Woessner and others, 1981; McClymonds, 1982).

Coal beds are important aquifers in southeastern Montana because generally the beds can be more laterally continuous than sandstone. Thus, domestic and livestock wells typically are completed in coal. Springs that issue from coal aquifers in outcrop and subcrop areas can provide water for livestock and wildlife and base flow to streams (Woessner and others, 1981; McClymonds, 1982; Wheaton and Donato, 2004). Springs are present throughout the Powder River Basin but are particularly abundant along contacts at the base of clinker zones and coal outcrops. In the Powder River Basin, springs have an average density of at least one spring per 5 mi^2 (Kennelly and Donato, 2001).

Between each of the five coal aquifers examined for this study, stratigraphic intervals of less permeable interbedded sandstone, siltstone, shale, and coal (fig. 2) can range from about 30 to as much as 400 ft thick. Because these intervals can be less permeable, they probably act, to some extent, as confining or leaky-confining layers depending on the location and permeability of the beds (Woessner and others, 1981). The underlying Lebo Shale Member is reported to be a limited source of water in the Powder River Basin but only where local coarse-grained channel sandstones also are found (Lewis and Roberts, 1978).

Previous Investigations

The most comprehensive reports on the hydrology and coal resources of the Northern Cheyenne Indian Reservation were the product of a 3-year research project that assessed the potential effects of surface mining of coal on groundwater and surface-water resources (Woessner and others, 1981; and the companion data report by Andrews and others, 1981). The researchers collected and published a large variety and volume of data, including information about the extent and thickness of coal beds; water-level and water-quality data from test holes, monitoring wells, and domestic wells; and hydraulic properties of coal, clinker, and sandstone aquifers.

Notable reports containing general information on geology or hydrology of the reservation include a report on geology and groundwater resources of Rosebud County (Renick, 1929), a report on geology and water-yielding characteristics of rocks in the northern Powder River Basin (Lewis and Roberts, 1978), and a map of geology and distribution of clinker in the northern Powder River Basin (Heffern and others, 1993). A report on the general geology, water quality, and occurrence of groundwater in the reservation was published by Hopkins (1973). The Montana Bureau of Mines and Geology (MBMG) published geologic maps of the Lame Deer 30×60 minute quadrangle (Vuke and others, 2001a) and the Birney 30×60 minute quadrangle (Vuke and others, 2001b), which encompass the entire study area.

Many reports by the USGS, Bureau of Land Management (BLM), and the MBMG contain detailed information about coal deposits on lands adjacent to the reservation and much of the Powder River Basin of southeastern Montana (Baker, 1929; Bass, 1932; Warren, 1960; Matson and Blumer, 1973; Mapel, 1976; Robinson and Culbertson, 1984; Derkey, 1986; Culbertson and Saperstone, 1987a, b; Biewick and McLellan, 1990, and Gruber, 1990). Coal data from wells and drill holes in the area were published in several reports (U.S. Geological Survey and Montana Bureau of Mines and Geology, 1980; Hansen and Culbertson, 1985; Culbertson, 1987; Wheaton and Donato, 2004).

Location-Numbering System

For this report, location numbers are used to identify wells according to their geographic position within the rectangular grid system used for the subdivision of lands (fig. 3). The location number consists of as many as 14 characters and is assigned according to the location of a well within a given township, range, and section. The first three characters (for example, 05S) specify the position of a well in a township south (S) of the Montana Base Line, whereas the next three characters (40E) specify its position east (E) of the Montana Principal Meridian. The next two characters

(31) are the section number; the next three to four characters (here, BDCC) designate the quarter section (160-acre tract), the quarter-quarter section (40-acre tract), the quarter-quarter-quarter section (10-acre tract), and the quarter-quarter-quarter-quarter section (2.5-acre tract), respectively, in which the well is located. These four subdivisions of the section are designated A, B, C, and D in a counter-clockwise direction, beginning in the northeastern quadrant. The last two characters (01) specify a sequence number to distinguish between multiple wells in a single tract. For example, as shown in figure 3, well 05S40E31BDCC01 is the first well inventoried in the SW¼ of the SW¼ of the SE¼ of the NW¼ of sec. 31, T. 1 S., R. 40 E.

Wells have been assigned other identifiers because data from these sites have been used in other investigations. For clarity and continuity, these site numbers are presented in this report as well.

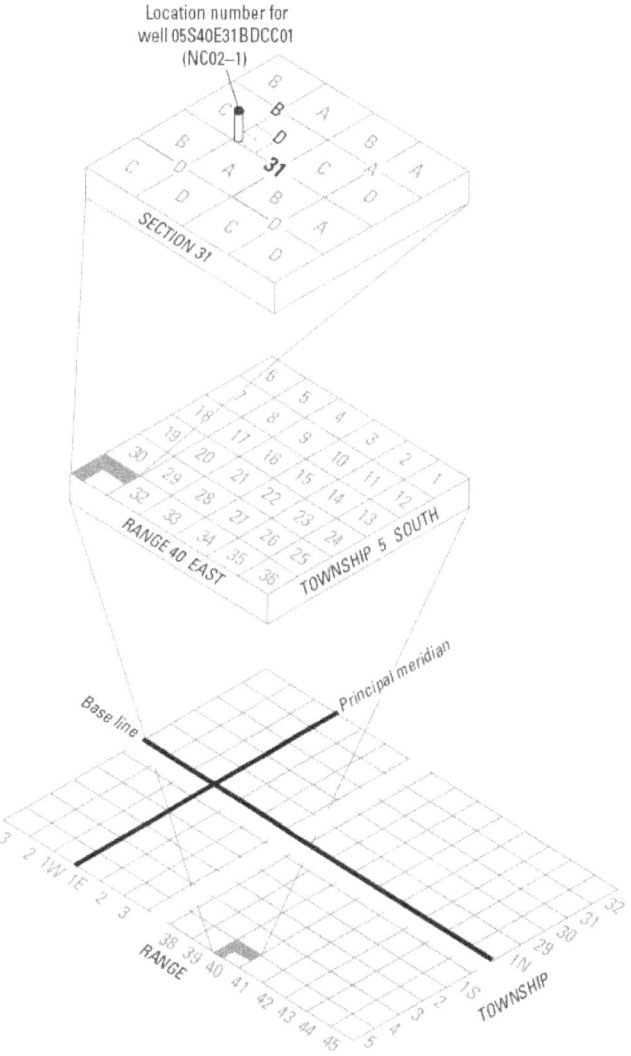

Figure 3.   Location-numbering system and site number for wells in and near the study area, southeastern Montana.

Methods of Investigation

Hydrogeologic data were compiled from various sources including the USGS National Water Information System—Groundwater Information and Data (NWIS) database, MBMG Ground-Water Information Center (GWIC) database, Woessner and others, (1981), and various coal-resource investigations of areas primarily outside the reservation. Proprietary drill-hole data from coal, oil, and gas wells were obtained from the Northern Cheyenne Tribe.

Estimation of Extent, Thickness, and Volume of Coal Beds

For each of the five coal beds evaluated in this study, the extent, thickness, and volume of the coal beds were estimated by using available well, drill-hole, and outcrop data in the reservation and nearby areas (fig. 1). Coal beds throughout the study area were indentified manually by considering correlations determined in previous studies, structural relations, and observations of geologists and Tribal officials. Altitude of the top and bottom of each coal bed then was used for correlation, to the extent possible, across the study area. Coal-bed thickness could be determined only from well and drill-hole data. In some instances, thickness data were not available where coal beds crop out; these outcrops typically are obscured by clinker and overlying sediments that have subsided into the burned area that formed the clinker (Heffern and Coates, 2004; Heffern and others, 2007). Geologic information and correlation of coal beds in the study area relied primarily on geologic information from McKay (1976a, b), Woessner and others (1981), Andrews and others (1981), Culbertson (1987), Culbertson and Saperstone (1987a, b), Biewick and McLellan (1990), McLellan and others (1990), and Heffern and others (1993), and from proprietary data. Additionally, to estimate the volume of the Flowers-Goodale coal, thickness information from Biewick and McLellan (1990) were used to supplement other data compiled for this report.

Data for each well and drill hole included location (latitude and longitude), altitude of tops and bottoms, and thickness (in feet) for each of the five coal beds. These data were compiled into a single geodatabase and processed by computer interpolation methods to estimate the volume of each coal bed. This geodatabase stored both tabular and spatial data in a format that was used internally and externally with software applications for a geographic information system (GIS). This approach provided a flexible platform for querying information about data in the geodatabase in multiple software platforms. ArcGIS, ArcScene 9.3.1 [Environmental Systems Research Institute (ESRI), 2010], and Microsoft Access 2007 (Microsoft Corporation, 2010) were used to manage data and estimate the volume of each coal bed.

By using ArcGIS 9.3.1 tools, the top surface of each coal bed was interpolated between data points by using an inverse distance weighting (IDW) method. This method of interpolation estimates and assigns thickness values to locations (cells) by averaging the weighted values of known thickness at a data point (well and drill-hole data) near each processing cell. The closer a data point is to the center of the cell being estimated, the more influence, or weight, it has in the averaging process (Environmental Systems Research Institute, 2010).

Raster-based surfaces representing the thickness of each coal bed were developed from point data and manually contoured lines of equal thickness (4-ft contour intervals); the contour data were subsequently digitized and used as input data. Raster-based surfaces representing the coal thickness were created by using the "Topo to Raster" tool in ArcGIS 9.3.1, which is an interpolation method specifically designed to create raster surfaces and hydrologically corrected digital-elevation models (DEMs). The interpolation procedure uses commonly available data, such as point, line, and polygon features and the known characteristics of altitude of the raster-based surfaces. The method uses iterative, finite-difference interpolation that is optimized to have the computational efficiency of local interpolation methods (such as IDW interpolation) without losing the surface continuity of other interpolation methods, such as kriging and spline (Environmental Systems Research Institute, 2010).

By using both of the raster-based surfaces created for the top elevation and the thickness for each of the five coal beds, the additive "Map Algebra" functions in ArcGIS 9.3.1 were then used to calculate the bottom altitude for each coal bed. All of the surfaces created (top, bottom, and thickness) were reviewed, inconsistencies and errors were identified, and surfaces were re-created through an iterative process.

Last, by using the raster-based thickness surface for each coal bed, volume estimates were calculated by using the "3D Analyst" extension in ArcGIS 9.3.1. The "Surface Volume" tool was used to calculate the volume of the raster-based surface relative to a given base height or reference plane. The tool was used to calculate the volume as cubic meters (subsequently calculated as acre-feet) between the plane and the top of the surface. Data were reviewed and stored in the final geodatabase.

The results of the computer interpolations used for this study are simplified representations of very complex physical conditions. One important assumption used to estimate the volume of the coal beds was that each coal was laterally continuous for an arbitrary distance of 1,320 ft (0.25 mi) beyond each data point. Additionally, the volume of each coal bed was only estimated to 1,320 ft beyond the 4-ft contour because it was assumed that coal aquifers that were less than 4 ft thick were unlikely to yield substantial amounts of water. Furthermore, another important assumption used to estimate the volume of each coal bed was that the coal beds were presumed to be continuous and uniform between data points where the volume was estimated. However, these coals are not necessarily continuous and uniform (fig. 4). Typically, coal beds can change in relatively short distances by (1) splitting, (2) merging, (3) pinching out, (4) differential compaction, and (5) fault displacement (Flores and others, 2010). For example, Woessner and others (1981, fig. 5. 13) report that in the western part of T. 3 S., R. 43 E., the Knobloch coal bed splits to form the upper Sawyer and lower Knobloch coal beds. These researchers also report many other instances in which this and other coal beds split, merge, or pinch out in the reservation and, thus, most likely in the study area.

Estimation of the Volume of Water in Aquifers

The volume of water in each of the five coal aquifers was estimated by assuming that each coal bed was fully saturated (thickness of the coal bed was equal to thickness of the coal aquifer) and by using reported aquifer characteristics for coal beds in areas of Montana, North Dakota, and Wyoming. Water-level data (period of record from 1976–2009, table 1) also were compiled to determine whether coal beds were unconfined or confined and to determine the percentage of saturation of each coal bed. Based on the compilation, estimates of the volume of water in the aquifers were determined for 25, 50, 75, or 100 percent saturation of the coal bed when possible.

Aquifer Characteristics

Storage

The ability of unconfined coal aquifers to store water largely is related to the network of natural fractures (cleats) within the coal aquifer. Additionally, coal aquifers in the study area were determined to have a fracture system controlled by

bedding properties and perhaps to crustal movements (Morin, 2005). When an unconfined coal aquifer is dewatered, water stored within the cleats is drained by gravity; cleats within coal aquifers provide the permeability for fluid flow (Purl and others, 1991). The volume of water derived by gravity drainage of a saturated coal aquifer (specific yield or effective porosity) is no larger than the fracture porosity of the coal aquifer (Rehm and others, 1980).

In an unconfined aquifer, the volume of water derived from the expansion of water and the compression of the aquifer is negligible (Heath, 1983). Thus, for unconfined coal aquifers in the study area, the volume of water that can be obtained by gravity drainage can be estimated by using the equation (modified from Heath 1983):

$$V_w = S_y \times V_a, \tag{1}$$

where

V_w is the volume of water drained from the coal aquifer by gravity, in acre-feet;

S_y is the specific yield, dimensionless;

V_a is the volume of the aquifer, in acre-feet.

Various researchers have examined the hydraulic properties of coal aquifers in areas of Montana, North Dakota, and Wyoming and have measured or estimated the specific yield of coal aquifers to range from about 0.3 to 3 percent (Groenewold and others, 1979; Rehm and others, 1980; Woessner and others, 1981; Davis, 1984). Thus, fracture porosity in the coal aquifers investigated in this study was assumed to range from 0.3 to 3 percent and average about 1 percent by volume (specific yield of 0.01). A specific yield of 0.01 was used to estimate the volume of water in each of the five coal aquifers where the volume of bed was estimated by computer interpolation.

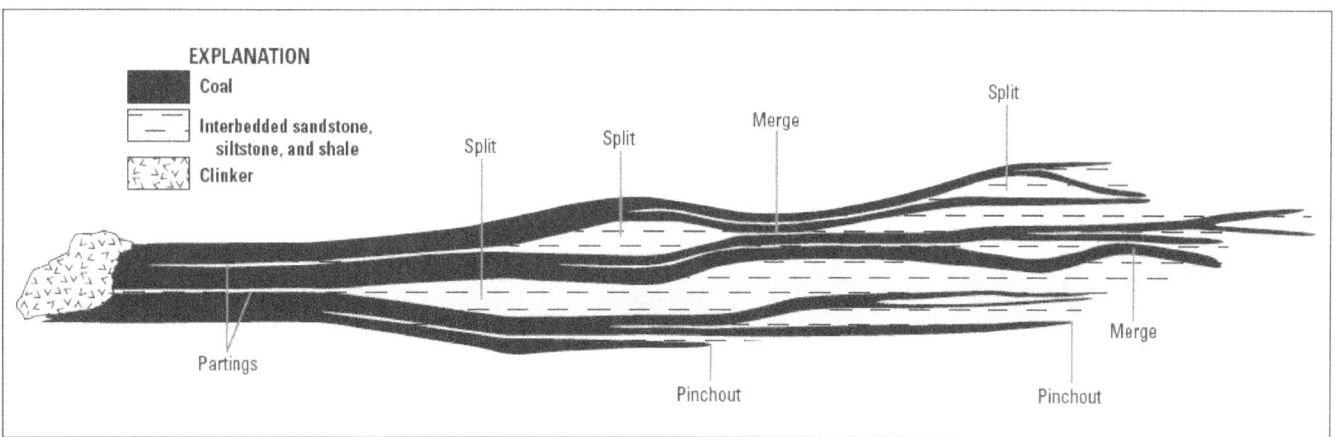

Figure 4. Conceptual model showing complex thinning, splitting, merging, and pinching out of coal beds (modifed from Flores and others, 2010).

Table 1.   Hydrogeologic data for selected wells completed in five coal beds in and near the study area, southeastern Montana.

[Location-numbering system described in text. Latitude and longitude are referenced to the North American Datum of 1983 (NAD 83). All altitudes are reported in feet above National Geodetic Vertical Datum of 1929 (NGVD 29). Altitudes of static-water levels are rounded to the nearest foot. Abbreviations: CBM, coal-bed methane; C, confined; E, electric tape; ft, feet; gal/min, gallons per minute; GWIC, Ground Water Information Center (Montana Bureau of Mines and Geology); M or MBMG, Montana Bureau of Mines and Geology; S, source of data, U.S. Geological Survey; SO, sounder; TOC, top of casing; UC, unconfined; USGS, U.S. Geological Survey; W, data from Woesnner and others (1981). Symbol: – –, no data or no remarks]

Location number[1]	Site number	Latitude	Longitude	Altitude of land surface or top of the casing[2]	Well depth, in feet, below land surface	Static-water level, in feet, below land surface or top of the casing[2]	Altitude of static-water level, below measuring point[2]	Altitude of top coal aquifer, below land surface	Percent saturation of coal bed the at the time of water-level measurement
		in degrees, minutes, seconds							
		\multicolumn{10}{c}{Canyon coal}							
05S40E13ADAB01	NC02–4	452429	1064354	3,940	326	– –	– –	[4]3,916	– –
05S40E31BDCC01	NC02–1	452139	1065049	4,440	655	– –	– –	4,079	0
		\multicolumn{10}{c}{Wall coal}							
05S40E13ADAB01	NC02–4	452429	1064354	3,940	326	200.07	3,740	3,636 C	100
05S40E31BDCC01	NC02–1	452139	1065049	4,440	655	624.7	3,815	3,828 UC	76
06S40E01CDDC01	PDC–33	452021	1064652	3,940	250	96.15	3,844	3,702 C	100
06S40E02DBDA01	PDC–6	452040	1064744	4,030	278	270.32	3,760	3,803 UC	6
06S41E06ABBC01	DH79–108	452101	1064526	4,020	348	273.29	3,747	3,681 C	100
		\multicolumn{10}{c}{Pawnee coal}							
05S41E14BDCD01	NC02–6	452408	1063833	3,510	356	– –	– –	[5]3,402 C	5,100
05S41E17ADBD01	NC02–3	452416	1064132	3,740	348	182.25	3,558	3,418 C	100
		\multicolumn{10}{c}{Knobloch coal}							
02S43E35CBC01	NCRP 27	453702	1061630	3,380	283	254.04	3,126	3,183 UC	28
03S43E01BDAA01	NCRP 31C	453134	1062245	3,220	176	126.38	3,094	3,106 UC	73
03S43E15AAAA01	NCRP 29A	453457	1062401	3,275	260	38.90	3,236	3,097 C	100
03S43E23BCC01	NCRP 30A	453342	1062009	3,080	124	85.15	2,995	2,989 C	100
04S42E36BDAA01	NCRP 11A	452707	1062931	3,302	265	138.08	3,164	3,052 C	100
05S41E14BDCD01	NC02–6	452408	1063824	3,510	356	238.94	3,271	3,174 C	100
05S42E14ADDC01	NC02–2	452412	1063018	3,220	394	– –	– –	3,075 UC	– –
05S42E16CCAB01	NC02–5	452355	1063339	3,400	370	262.58	3,137	3,055 C	100
05S42E28DDAC02	CBM01–8KC	452208	1063252	3,262	208	158.00	3,104	3,070 C	100
05S43E07DBBA01	NC97–3	452500	1062812	3,115	240	– –	– –	– –	– –
		\multicolumn{10}{c}{Flowers-Goodale coal}							
05S42E14ADDC02	NC02–2	452412	1063018	3,220	394	107.40	3,113	2,856 C	100
05S42E28DDAC01	CBM02–8FG	452208	1063252	3,261	480	101.59	3,159	2,801 C	100
05S43E07CCDC01	NC05–1	452438	1062839	3,170	750	Flowing; static-water level not available	– –	2,856 C	100
05S43E07CCDC02	NC05–2	452438	1062840	3,170	348	49.25	3,121	2,865 C	100

Table 1. Hydrogeologic data for selected wells completed in five coal beds in and near the study area, southeastern Montana.—Continued

[Location-numbering system described in text. Latitude and longitude are referenced to the North American Datum of 1983 (NAD 83). All altitudes are reported in feet above National Geodetic Vertical Datum of 1929 (NGVD 29). Altitudes of static-water levels are rounded to the nearest foot. Abbreviations: CBM, coal-bed methane; C, confined; E, electric tape; ft, feet; gal/min, gallons per minute; GWIC, Ground Water Information Center (Montana Bureau of Mines and Geology); M or MBMG, Montana Bureau of Mines and Geology; S, source of data, U.S. Geological Survey; SO, sounder; TOC, top of casing; UC, unconfined; USGS, U.S. Geological Survey; W, data from Woesnner and others (1981). Symbol: – –, no data or no remarks]

Location number[1]	Method of water-level measurement	Date of water-level measurement	Source of data	Remarks[3]
Canyon coal				
05S40E13ADAB01	– –	– –	S	Geologist's log reported "from 24–44 ft, Coal (Canyon)." Well was completed in Wall coal.
05S40E31BDCC01	– –	– –	S	USGS geologist's log reported "from 361–382 ft, Coal, black, dry (Canyon)."
Wall coal				
05S40E13ADAB01	E	11/03/09	S	– –
05S40E31BDCC01	E	06/06/05	S	– –
06S40E01CDDC01	S	03/06/82	S	Geologist's log reported 7 ft of parting between the upper Wall coal (47 ft thick) and lower Wall coal (10 ft thick). The packer was set at the top of the lower Wall coal and the well was perforated in this same bed.
06S40E02DBDA01	E	03/06/82	S	Geologist's log reported 46 ft of Wall coal.
06S41E06ABBC01	E	03/06/82	S	Most likely completed in the upper Wall coal.
Pawnee coal				
05S41E14BDCD01	– –	– –	S	Geologist's log reported "from 108–130, Coal (producing 15 gal/min, Pawnee)." Well was completed in Knobloch coal. Nearby test hole drilled to 740 ft. Combined depth shown on fig. 6 for NC02–6. Drilled to top of Lebo Shale Member of the Tertiary Fort Union Formation.
05S41E17ADBD01	E	11/03/09	S	Geologist's log reported "from 322–346, Coal (Pawnee); hole producing more than 20 gal/min."
Knobloch coal				
02S43E35CBC01	W	10/20/76	W	– –
03S43E01BDAA01	W	10/20/76	W	– –
03S43E15AAAA01	E	11/02/09	S	– –
03S43E23BCC01	W	10/20/76	W	– –
04S42E36BDAA01	W	10/20/76	W	– –
05S41E14BDCD01	E	11/03/09	S	Nearby test hole drilled to 740 ft. Combined depth shown on fig. 6 for NC02–6. Drilled to top of Lebo Shale Member.
05S42E14ADDC01	– –	– –	S	Geologist's log reported "145–166 ft, coal, black moist (upper Knobloch), saturated in lower part." Drilled to Flowers-Goodale coal.
05S42E16CCAB01	E	11/03/09	S	– –
05S42E28DDAC02	SO	08/28/09	M	MBMG other identifier: GWIC 203697
05S43E07DBBA01	– –	– –	S	Well not completed in Knobloch coal; geologist's log used for correlation (fig. 6).
Flowers-Goodale coal				
05S42E14ADDC02	E	11/03/09	S	– –
05S42E28DDAC01	SO	08/28/09	M	MBMG CBM monitoring well and other identifier: GWIC 203701; MBMG geologist's log reported "Brewster-Arnold damp at base."
05S43E07CCDC01	– –	– –	S	Geologist's log reported "314–340 ft, coal (Flowers-Goodale); hole producing about 30 gal/min." Drilled to determine top of Lebo Shale Member of the Tertiary Fort Union Formation.
05S43E07CCDC02	E	11/03/09	S	– –

[1]Some locations from Woessner and others (1981) were not field checked by the USGS. For these sites, latitude and longitudes were estimated from the Montana State Library—Natural Resouce Information Sysytem (*http //maps2.nris.mt.gov/topofinder1/subsection.asp*) that converts public-land survey information to latitude and longitude. Some of these wells were field checked in summer 2010 and were then assigned more accurate latitudes and longitudes.

[2]Static-water level data from Woessner and others (1981) are reported from the top of the casing; all other static-water level data are reported from land surface.

[3]All logs are from the USGS (unless noted otherwise) and on file at USGS Montana Water Science Center, Helena, Mont.

[4]The Canyon coal at this well was assumed to be dry because the top of the coal was only 24 ft below land surface, and the same coal was dry at nearby well NC02–1.

[5]The Pawnee coal at this well was assumed to be fully saturated and under confined conditions because a geologist's log reported that the well was producing 15 gal/min at that interval; the same aquifer was under confined conditions at nearby well NC02–3.

Water storage in confined coal aquifers (storage coefficient or storativity, Freeze and Cherry, 1979) is related to properties of both cleats and compressible storage in the unfractured coal matrix. When hydraulic head is reduced in a confined coal aquifer, some water is released from storage through the expansion of water and compression of the aquifer material. Release of water from the cleat system is much faster than from the coal matrix because of the large difference in hydraulic conductivity between the cleats and coal matrix. However, compressible storage in the coal matrix is generally substantially greater than that in the cleats (Weeks, 2005). A storage coefficient calculated from a multiple-well aquifer test in the Flowers-Goodale coal (25 ft thick) near Birney Day School was 1.4×10^{-4} (Weeks, 2005). The corresponding specific storage (defined as the volume of water that a unit volume of confined aquifer releases from storage under a unit decrease in hydraulic head, Freeze and Cherry, 1979) was 5.6×10^{-6}/ft. The specific storage of the Flowers-Goodale coal compares favorably with a specific storage of 8×10^{-6}/ft for the Anderson coal and 2×10^{-5}/ft for the Sawyer-A coal in the Powder River Basin of Montana (Stoner, 1981). Additionally storage coefficients for coal aquifers in the Decker area [the Anderson, Dietz 1, and Dietz 2 coal beds of Baker (1929)] ranged from about 1.0×10^{-5} and 3.0×10^{-5} (Van Voast and Hedges, 1975). The storage coefficients of most confined aquifers range from 1.0×10^{-3} to 1.0×10^{-5} and is about 1.0×10^{-6} per foot of thickness (Lohman, 1979).

If the water level in a confined coal aquifer decreases below the top of the bed, then the aquifer becomes unconfined and the volume of water is then derived by gravity drainage; the volume of water can then be estimated by using equation 1. For example, by using a specific yield of 0.01, a coal aquifer that was 10-ft thick when dewatered by gravity drainage, would yield 0.1 acre-feet (acre-ft) of water per acre from a 10-ft decrease in hydraulic head. With a specific storage of 1.0×10^{-6}/ft, a confined coal aquifer that was 10-ft thick would yield 0.0001 acre-ft of water per acre from a 10-ft decrease in hydraulic head. This calculation shows that the volume of water stored in confined coal aquifers by compressible storage in the unfractured coal matrix typically would be at least three orders of magnitude smaller than the volume of water derived by gravity drainage of saturated coal. However, Freeze and Cherry (1979) and Lohman (1979) noted that large hydraulic head changes over extensive areas can produce substantial volumes of water from confined aquifers.

For this report, the estimated volume of water released from confined storage from coal aquifers also was assumed to be small when compared with the estimated volume of water derived by gravity drainage. Therefore, despite the fact that all five coal aquifers may exist in both unconfined and confined conditions in the study area, only the volume of water that could be drained by gravity from storage under unconfined conditions was included in estimates of the volume of water in each coal aquifer.

Hydraulic Conductivity and Transmissivity

The hydraulic conductivity of rock or soil is a measure of its ability to transmit water and typically is reported in units of foot per day (ft/d). Transmissivity also is used as a measure of how rock or soil can transmit water and is reported in feet squared per day (ft^2/d). Hydraulic conductivity values of coal, sandstone, and interbedded siltstone, clay, and shale of the Fort Union Formation are described here to show the relatively large hydraulic conductivity and transmissivity of the coal aquifers as compared with the other sedimentary rocks that make up the formation. The larger hydraulic conductivity and transmissivity of the laterally extensive coal aquifers makes these aquifers favorable conduits for groundwater flow.

The hydraulic conductivity of the Flowers-Goodale coal near Birney Day School was 4.3 ft/d (Weeks, 2005). Slagle and others (1985) reported that hydraulic conductivity for the Anderson coal ranged from 0.34 to 6.5 ft/d and transmissivity ranged from 11 to 223 ft^2/d. McClymonds (1982) reported that hydraulic conductivity for the Wall coal (just south of the study area) ranged from 0.05 to 2.4 ft/d and transmissivity from the same aquifer tests ranged from 2.5 to 65 ft^2/d. Woessner and others (1981) reported that hydraulic conductivity for the Knobloch coal ranged from 0.9 to 2.9 ft/d. Last, Van Voast and Hedges (1975) reported that hydraulic conductivity for coal beds in the Decker area ranged from 0.5 to 19 ft/d and transmissivity ranged from 5 to 270 ft^2/d.

For hydraulic conductivity, Wheaton and Metesh (2002) determined a geometric mean value of 1.1 ft/d (from 370 values) for coal aquifers in the Powder River Basin of Montana, with standard deviations of 13 and 0.098 ft/d (plus or minus one standard deviation, respectively). The values of hydraulic conductivity for coal aquifers in the study area and in the Powder River Basin are somewhat larger than the geometric mean of hydraulic conductivity of 0.18 ft/d, with standard deviations of 2.1 and 0.015 ft/d (plus or minus one standard deviation, respectively) reported for sandstone in the Powder River Basin of Montana (Wheaton and Metesh, 2002). In contrast, Rehm and others (1980) determined a geometric mean for hydraulic conductivity of 0.007 ft/d (from 63 values) for Paleocene silt, clay, and shale in Montana, Wyoming, and North Dakota.

Water-Level Data

Water-level data were compiled to estimate the volume of water in each of the five coal beds and to determine whether these coal beds were unconfined or confined. Water-level data from wells completed in these coal beds were scarce (table 1); thus, in all instances, geologists' logs were carefully reviewed to ensure that water-level data were determined for the specific coal bed for this study.

Estimates of the Volume of Water in Five Coal Aquifers

Estimates of the volume of water in the five coal aquifers were based on equation 1, a specific yield of 0.01, and the assumption that the five coal beds were unconfined. The volume of water was estimated by assuming that each coal bed was fully saturated (thickness of the coal bed was equal to the thickness of the coal aquifer, except for the Canyon coal). Additionally, estimates of the volume of water in the coal aquifers were determined only in areas where the volume of the coal beds could be estimated by computer interpolation. Woessner and others (1981) reported that the groundwater system within the Tongue River Member is complicated owing to complexly interstratified sandstone, shaley sandstone, siltstone, shale, and coal beds (figs. 2 and 4). Only conservative estimates of the volume of water in the Canyon, Wall, Pawnee, Knobloch, and Flowers-Goodale coal beds are presented because (1) the subsurface extent of the coal beds are not well defined, (2) in some instances, well and drill-hole data were widely spaced and not well distributed, (3) of the possibility that some coal beds split, merge, or pinch out laterally, and (4) water-level data for wells completed in the five aquifers were scarce.

Canyon Coal

The Canyon coal typically exists as erosional remnants in higher plateaus and can be highly dissected and, thus, has limited subsurface extent in the study area (figs. 5 and 6); the Canyon coal encompasses about 40,100 acres (about 63 mi^2). Where the volume of this coal could be estimated by computer interpolation, the Canyon coal averages about 26 ft thick (table 2, fig. 5). This coal bed splits and thins just north and west of the study area boundary (Woessner and others, 1981).

For unconfined conditions, the volume of water in the Canyon coal was estimated to range from about 10,400 acre-ft (75 percent saturated) to 3,450 acre-ft (25 percent saturated, table 2). Water-level data for the Canyon coal were unavailable, but one geologist's log (table 1) indicated that at well NC02–1 (fig. 5), the Canyon coal was dry. Because of its high topographic position (fig. 6), and because water-level data are unavailable to assess the amount of saturation of this coal, 100 percent saturation was not considered probable within the study area. The smaller estimates of water in the Canyon coal (75 to 25 percent saturation, table 2) are considered more reasonable. However, estimates of the volume of water in the Canyon coal might have large errors and need to be used with caution because the water-level data needed to define the volume of water were unavailable.

Because the Canyon coal is highly dissected in the study area, this coal most likely is composed of local flow systems with variable amounts of recharge and discharge at different locations. For example, Woessner and others (1981) noted

that in the southwest corner of the reservation, the base of the Canyon coal is the source of many springs. McClymonds (1982) also noted that just south of the study area near wells PDC–6 and PDC–33 (fig. 5), numerous seeps and springs issue from the Canyon coal. Additionally, McClymonds (1982) reported that this coal probably is a major contributor to increased streamflow in Prairie Dog Creek, which is a tributary to the Tongue River (fig. 1).

Wall Coal

The Wall coal, which has limited subsurface extent in the study area (fig. 7), encompasses about 54,800 acres (about 86 mi^2). Where the volume of this coal could be estimated by computer interpolation, the Wall coal averages about 44 ft thick (table 2, fig. 7). Near well NC02–1, the overlying Cook coal (fig. 2) was assumed to merge with the Wall coal (Woessner and others, 1981), which probably accounts for this coal's greater thickness along part of the southern study area boundary (fig. 7). Conversely, splits of the upper and lower Wall coal were noted in the study area as far west as T. 5 S., R. 40 E. (drill holes A–12 and A–13 of Culbertson, 1987). The thickness of the Wall coal is unknown in the eastern part of the study area where it exists near Black Eagle Fork. Woessner and others (1981, fig. 5.15) reported that the Wall coal thins (not shown on fig. 7) and eventually disintegrates into thin clinker and white clayey zones near Pawnee and Kelty Creeks.

For unconfined conditions, the volume of water in the Wall coal was estimated to range from about 14,200 acre-ft (100 percent saturated) to 3,560 acre-ft (25 percent saturated, table 2). Water-level data from two wells (NC02–1 and PDC–6) indicate that the Wall coal at these wells was unconfined, and water-level data from three wells (NC02–4, PDC–33, and DH79–108) indicate that this coal at these wells was confined (table 1, fig. 7). Just south of the study area, McClymonds (1982) determined that the Wall coal is composed of two beds with a parting of about 7 ft. The upper bed can be unconfined or dry, whereas water levels in a well completed in the lower bed indicated confined conditions (PDC–33; table 1, fig. 7). Because the Wall coal exists under unconfined and confined conditions and can be 100 percent saturated, the estimates of the volume of water in the Wall coal probably are reasonable.

Periodic water-level data have been collected from two wells (figs. 6 and 7) completed in the Wall coal that are located in the study area. These data show that water levels in well NC02–4 fluctuated less than 1.5 ft (periodic measurements from December 2002 through November 2010), whereas water levels in well NC02–1 fluctuated about 53 ft (periodic measurements from December 2002 through June 2005; water-level data for both wells were accessed at *http://nwis.waterdata.usgs.gov/mt/nwis/gwlevels*, April 11, 2011). The water level measured in well NC02–1 in December of 2002 (575.7 ft below land surface) is slightly higher than the rest of the measurements that ranged from 613.5 to 628.24 ft below land surface.

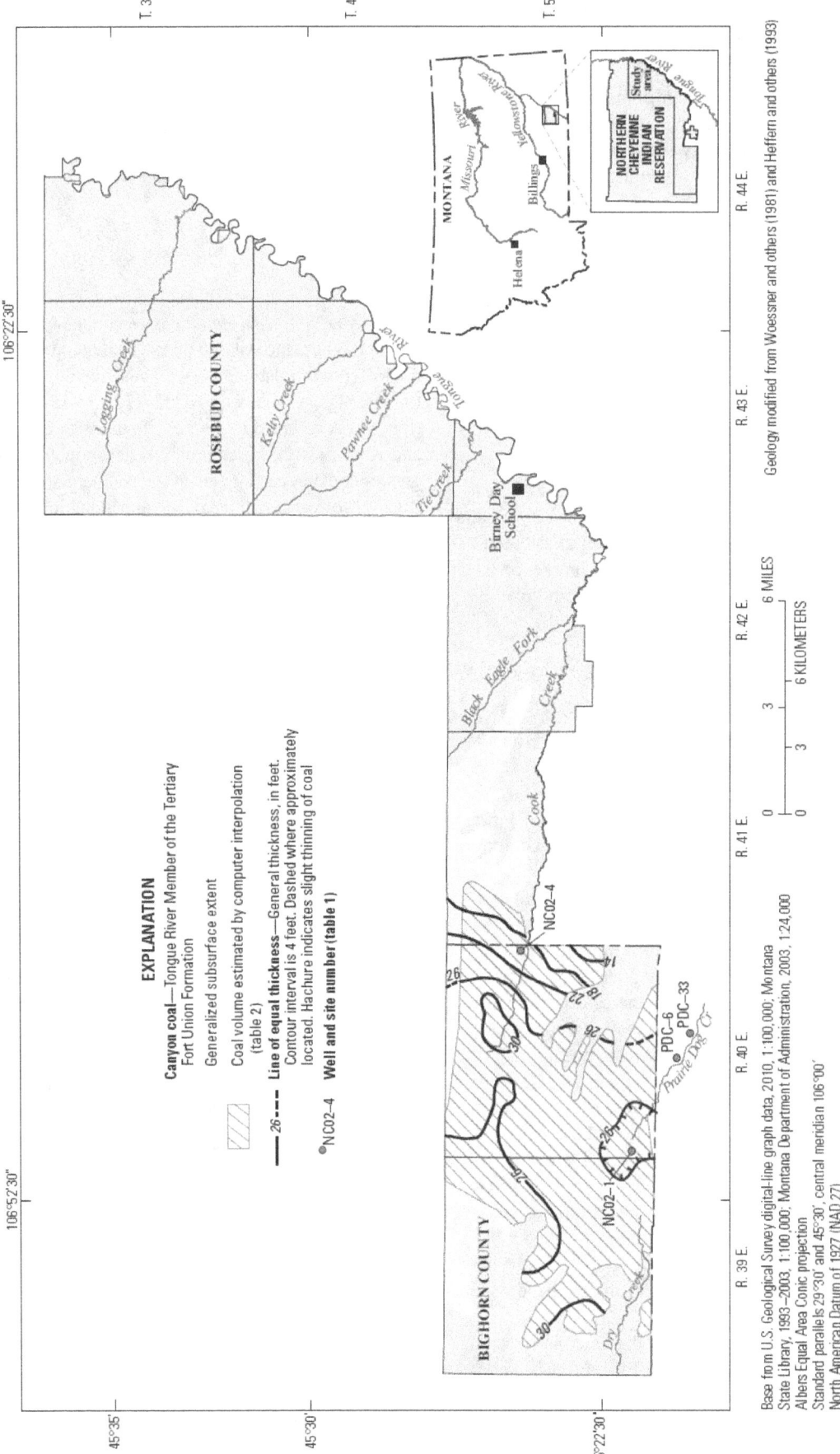

Figure 5. Generalized subsurface extent and thickness of the Canyon coal in the study area, southeastern Montana.

Pawnee Coal

The Pawnee coal, which is found in the subsurface throughout much of the study area (fig. 8), encompasses about 73,600 acres (about 115 mi^2). Where the volume of this coal could be estimated by computer interpolation, the Pawnee coal averages about 12 ft thick (table 2, fig. 8). This coal thins to the southwest and northeast and then thins and splits to the southeast (Woessner and others, 1981). Generally, the thickness of the Pawnee coal is unknown in the eastern part of the study area except near Pawnee and Kelty Creeks, where the coal ranges from about 6 to 10 ft thick.

For unconfined conditions, the volume of water in the Pawnee coal was estimated to range from about 9,440 acre-ft (100 percent saturated) to 2,360 acre-ft (25 percent saturated, table 2). Water-level data from one well (NC02–3, table 1, fig. 8) completed in the Pawnee coal indicated that the coal at this well was confined. Data from one geologist's log indicated that at well NC02–6, the Pawnee coal probably was fully saturated and might have been confined (table 1, fig. 8). However, estimates of the volume of water in the Pawnee coal might have large errors and need to be used with caution because the water-level data needed to define the volume of water were largely unavailable.

Periodic water-level data have been collected from one well (figs. 6 and 8) completed in the Pawnee coal that is located in the study area. These data show that water levels in well NC02–3 fluctuated less than 1.0 ft (periodic measurements from December 2002 through November 2010; *http://nwis.waterdata.usgs.gov/mt/nwis/gwlevels*, accessed April 11, 2011).

Knobloch Coal

The Knobloch coal, which is found in the subsurface throughout much of the study area (fig. 9), encompasses about 75,200 acres (about 117 mi^2). Where the volume of this coal could be estimated by computer interpolation, the Knobloch coal averages about 29 ft thick (table 2, fig. 9).

The Knobloch coal probably is one of the most complexly stratified geologic units in the study area. Heffern (1980), Woessner and others (1981), and Culbertson and Saperstone (1987b) reported that this coal is composed of a number of separate coal beds that merge near Ashland and can be as much as about 70 ft thick (McLellan and others, 1990). In the northeastern part of the study area near Logging Creek, part of the Knobloch coal is thought to split sharply from the main Knobloch coal bed to form the Sawyer coal (Woessner and others, 1981). North of Logging Creek, the Knobloch coal is composed of one main bed; south of Logging Creek, this coal bed splits into upper and lower beds. Near Birney Day School, the lower Knobloch bed splits again and the lower split is recognized as the Nance coal (Woessner and others, 1981; Culbertson and Saperstone, 1987b). Farther south, the Knobloch coal splits and thins (Heffern, 1980) and to the west is either absent or splits and thins and (or) also pinches out to the west (Culbertson and Saperstone, 1987b).

For unconfined conditions, the volume of water in the Knobloch coal was estimated to range from about 38,700 acre-ft (100 percent saturated) to 9,680 acre-ft (25 percent saturated, table 2). Water-level data from two wells (NCRP 27 and NCRP 31C) and information from one geologist's log (NC02–2) indicate that the Knobloch coal was unconfined at these wells, whereas water-level data from six wells (NCRP 29A, NCRP 30A, NCRP 11A, NC02–6, NC02–5, and CBM01–8KC) indicate this coal was confined at these wells (table 1, fig. 9). Water-levels in the two wells that indicate unconfined conditions show that this coal was at least 25 percent saturated when water levels were measured (table 1). Thus, estimates of the volume of water in the Knobloch coal probably are reasonable.

Water-level data have been collected in three wells (figs. 6 and 9) completed in the Knobloch coal that are located in the study area. Periodic water-level data indicate that water levels in well NCRP 29A fluctuated less than 11 ft (periodic measurements June 2007 through November 2010). By contrast, periodic water-level data indicate that water levels measured in wells NC02–5 and NC02–6 fluctuated less than about 2.5 ft (periodic measurements December 2002 through November 2010; water-level data for both wells were accessed at *http://nwis.waterdata.usgs.gov/mt/nwis/gwlevels*, April 11, 2011).

Flowers-Goodale Coal

The Flowers-Goodale coal, which is found in the subsurface throughout most of the study area (fig. 10), extends west and north of the study area throughout the rest of the Northern Cheyenne Indian Reservation (Biewick and McLellan, 1990). Biewick and McLellan (1990) also showed that in the southern part of T. 5 S., R. 39 E. and T. 5 S., R. 40 E. this coal is missing (fig. 10). The Flowers-Goodale coal encompasses about 132,400 acres (about 207 mi^2). Where the volume of this coal could be estimated by computer interpolation, the Flowers-Goodale coal averages about 17 ft thick (table 2, fig. 10) but is as much as 28 ft thick near Tie Creek (fig. 10). East and southeast of the Northern Cheyenne Indian Reservation, Biewick and McLellan (1990) showed that the Flowers-Goodale coal ranged from about 3 to 20 ft in thick and also showed that the bed is laterally continuous across the reservation boundary.

For unconfined conditions, the volume of water in the Flowers-Goodale coal was estimated to be about 35,800 acre-ft (100 percent saturated, table 2). Water-level data from three wells (NC02–2, CBM02–8FG, NC05–2) in and near the study area and information from one geologist's log (NC05–1) indicate that the Flowers-Goodale coal was confined at these wells (table 1, fig. 10). Also, because this

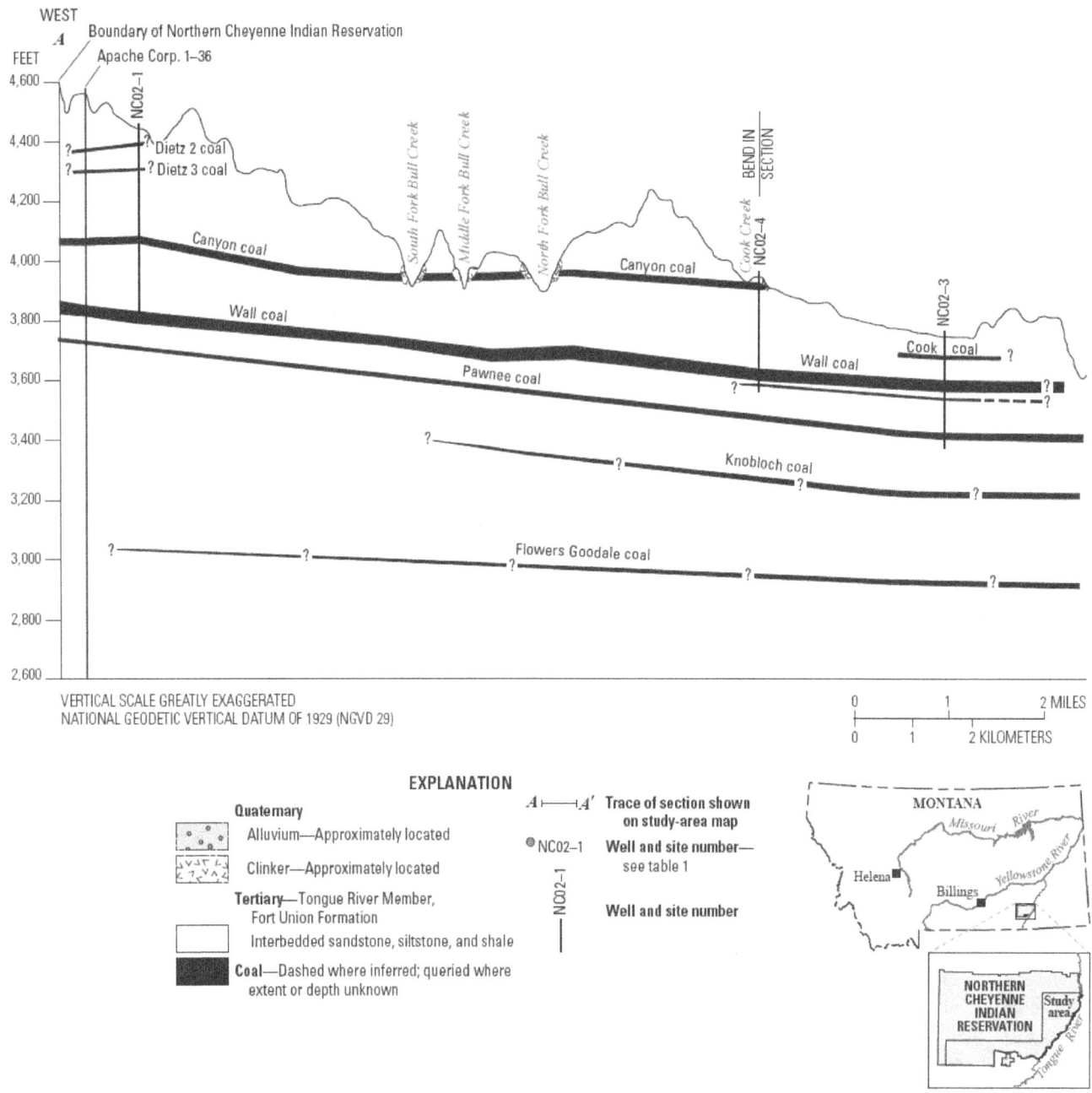

Figure 6 (above and facing page).   Generalized geologic section of the Tongue River Member of the Tertiary Fort Union Formation along the southern part of the Northern Cheyenne Indian Reservation, southeastern Montana.

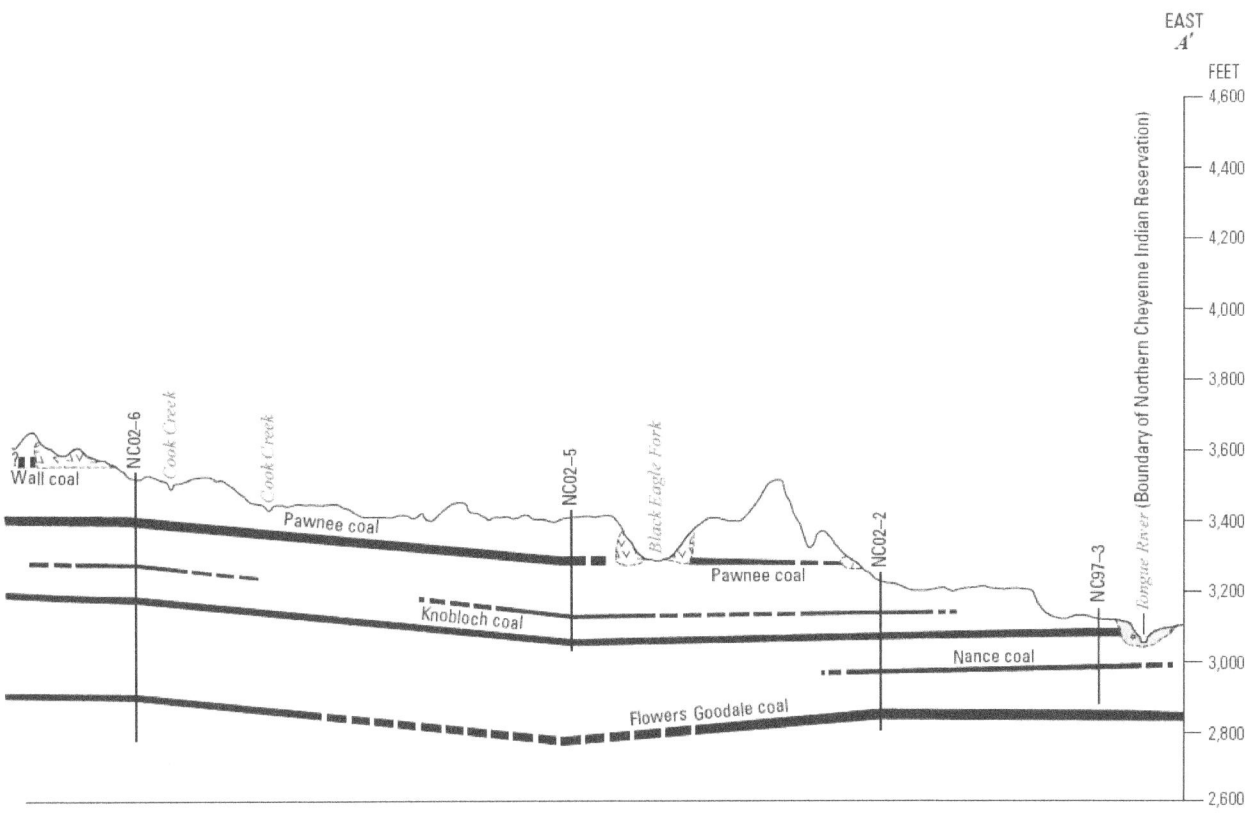

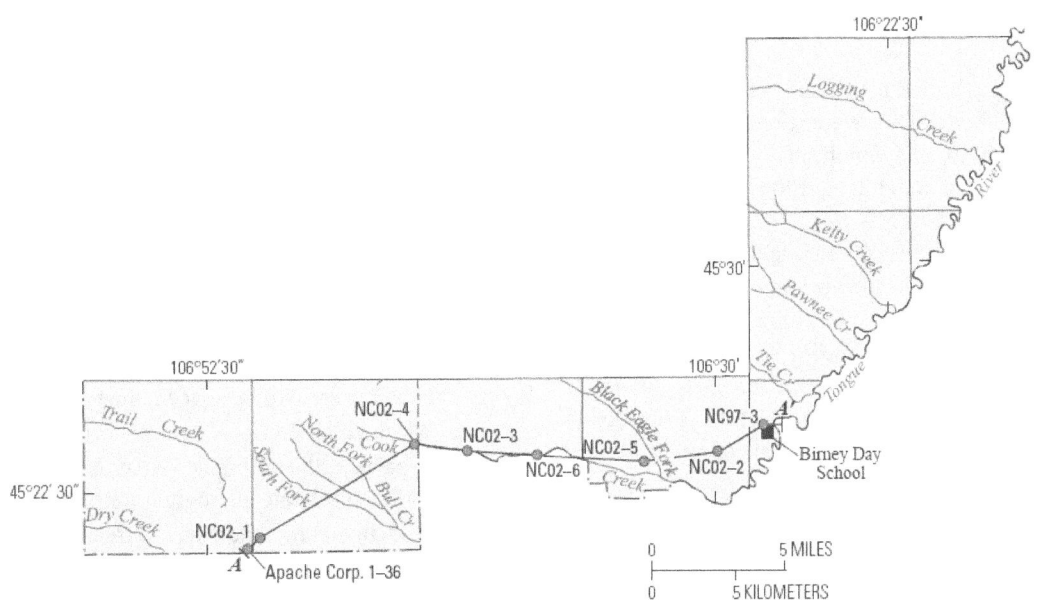

Table 2.   Range of estimates of the volume of water in five coal aquifers in the study area, Northern Cheyenne Indian Reservation, southeastern Montana.

[Coal beds (aquifers) are within the Tongue River Member of the Tertiary Fort Union Formation. Areas where the volume of the five coal beds was estimated are shown in figures 5 and 7–10. Symbol: --, no estimate]

Estimated subsurface extent of coal bed, in acres	Estimated subsurface extent of coal bed used for estimation of volume	Estimated average thickness of coal bed, in feet[1]	Estimated volume of coal bed, in acre-feet[2]	Estimated volume of water in coal aquifers, unconfined storage, in acre-feet[3]			
				Percent saturation			
				100	75	50	25
Canyon coal[4]							
40,100	31,200	26	1,380,700	--	10,400	6,900	3,450
Wall coal							
54,800	36,800	44	1,422,900	14,200	10,700	7,110	3,560
Pawnee coal[4]							
73,600	40,300	12	944,300	9,440	7,080	4,720	2,360
Knobloch coal							
[5]75,200	47,800	29	3,871,000	38,700	29,030	19,400	9,680
Flowers-Goodale coal							
132,400	73,600	17	3,578,000	35,800	[6]--	[6]--	[6]--

[1]Average thickness was calculated only for the area where the volume of the coal bed was estimated by computer interpolation.

[2]Methods for estimating volume of coal beds are described in the section "Estimation of Extent, Thickness, and Volume of Coal Beds" of this report.

[3]Methods for estimating volume of water in unconfined storage are described in the section "Storage" of this report. Estimated volumes are rounded to three significant figures.

[4]Estimates might have large errors because water-level data needed to define the volume of water do not exist.

[5]Generalized subsurface extent excludes that area where the extent of the Knobloch coal was uncertain (figure 9).

[6]Flowers-Goodale coal was assumed to be confined throughout the study area; data indicate that water levels can be as much as about 350 feet above the top of this bed (table 1).

coal generally is deeply buried in the study area (about 200 ft or more, fig. 6), the Flowers-Goodale coal was assumed to be confined throughout its extent in the study area. Thus, the estimates of the volume of water in the Flowers-Goodale coal probably are reasonable.

Water-level data have been collected in two wells (figs. 6 and 10) completed in the Flowers-Goodale coal that are located in the study area. Periodic and continuous water-level data from wells NC02–2 (period of record September 2002 through September 2010; U.S. Geological Survey, 2010b) and NC05–2 indicate that water levels in these wells have fluctuated less than about 2 ft (period of record June 2007 through November 2010, accessed at *http://nwis.waterdata.usgs.gov/mt/ nwis/gwlevels*, April 11, 2011).

Reliability and Uncertainty of Volume Estimates

As stated by Flores and others (2010), sufficient data are needed to accurately characterize coal-bed horizontal and vertical variability, which is highly complex both locally and regionally (fig. 4). Additionally, the distribution of wells or

drill holes is important because data from these sites is the ultimate control governing the reliability and uncertainty of any estimate of the volume of coal (Wood and others, 1983). Splitting, merging, pinching out, differential compaction, and fault displacement can further complicate correlation of coal beds and also can affect the reliability of the estimates of the volume of the coal beds between data points. Variations of coal thickness presented in this report might reflect variations influenced by some or all of these factors. Where data points are widely spaced, the reliability of estimates of the volume of coal beds is decreased. Likewise, reliable estimates of the volume of water in coal aquifers depend heavily on reliable geologic information, such as subsurface extent and thickness and correlation of coal beds. Additionally, reliable estimates of the volume of water in coal aquifers depend heavily on data from wells, such as up-to-date water-level and well-completion data and data about coal-aquifer characteristics. Because the data needed to define the volume of water were sparse, only conservative estimates of the volume of water in the five coal aquifers are presented in this report. These estimates need to be used with caution and mindfulness of the uncertainty associated with these estimates.

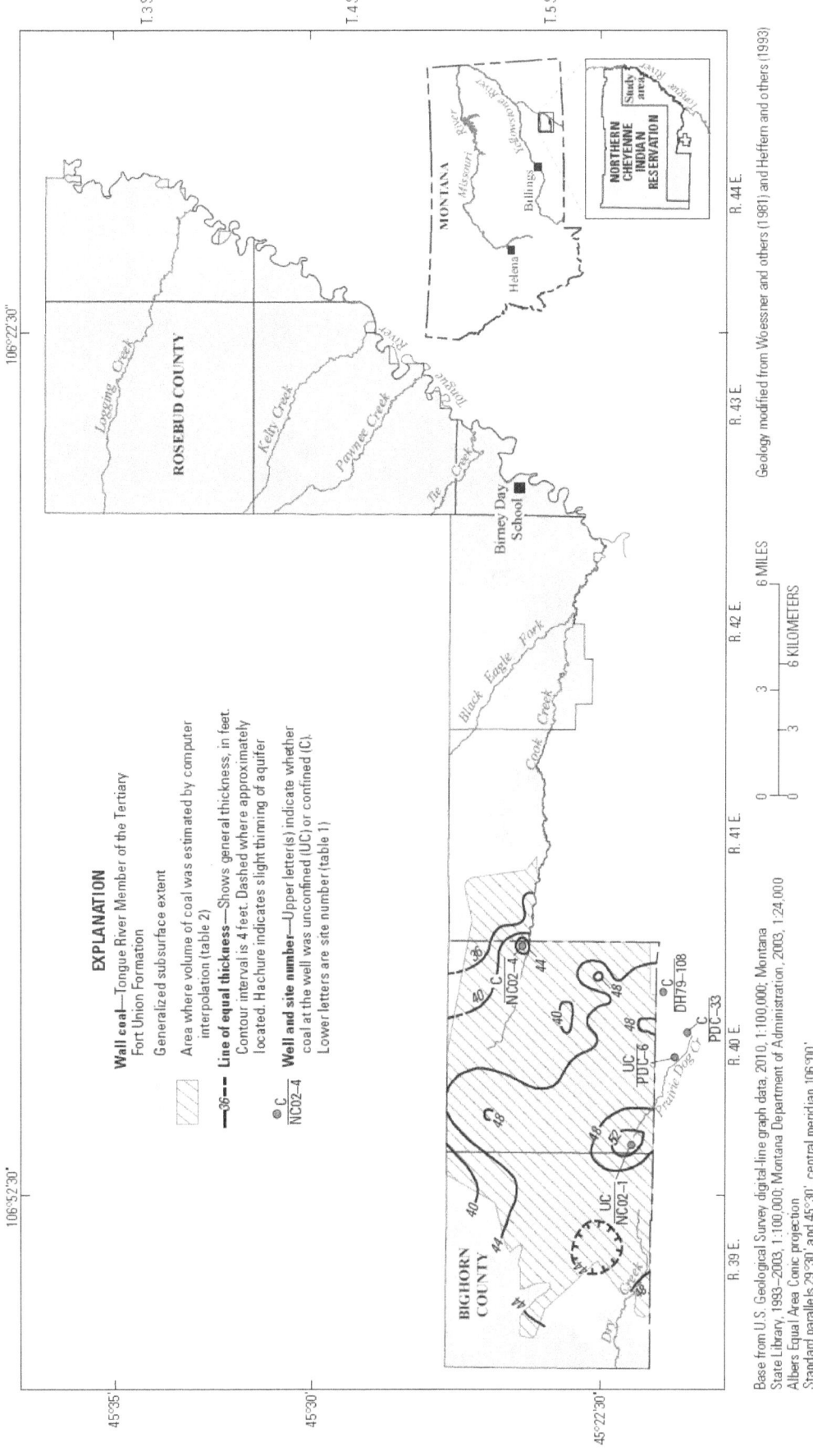

Figure 7. Generalized subsurface extent and thickness of the Wall coal in the study area, southeastern Montana.

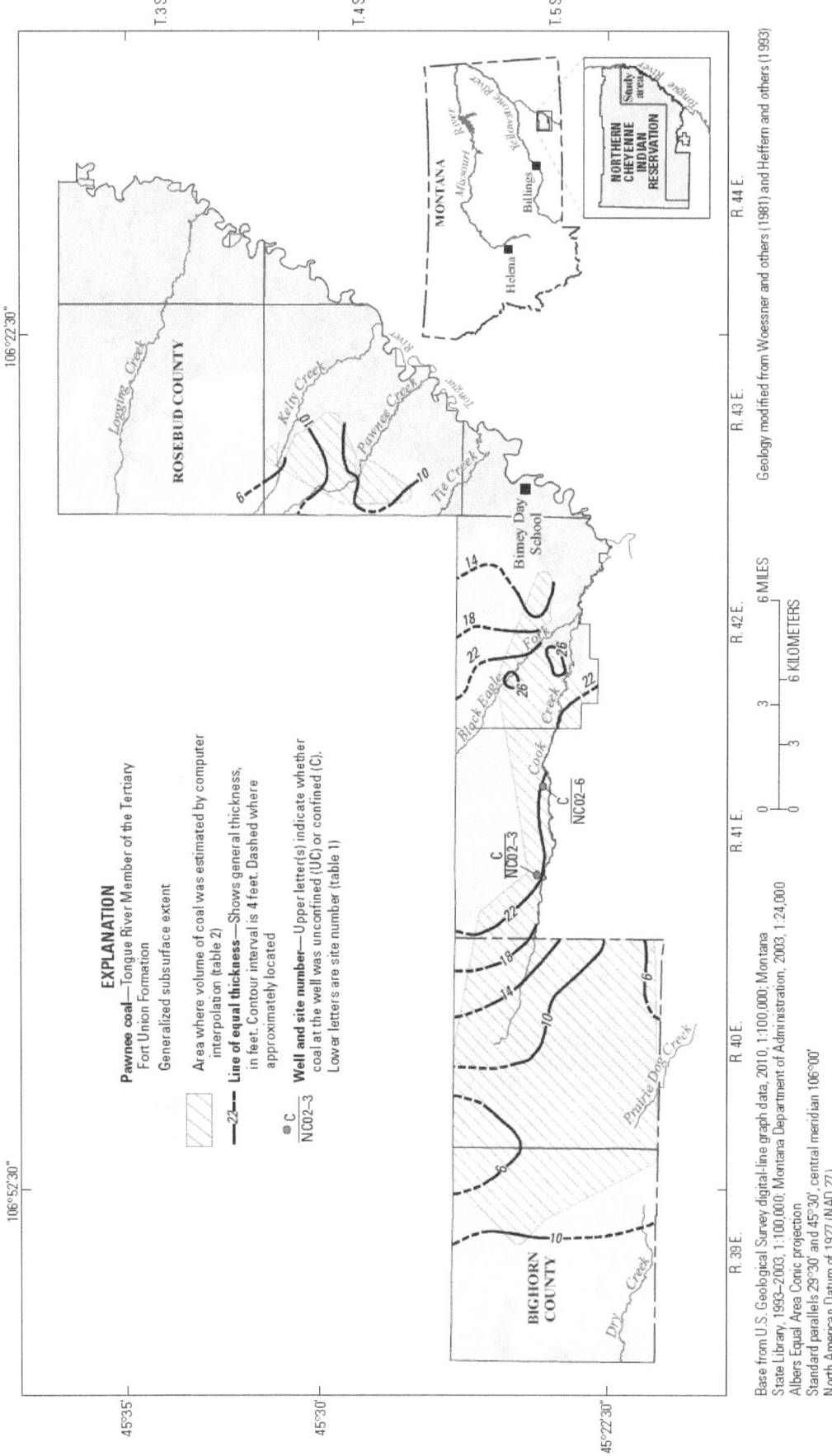

Figure 8.   Generalized subsurface extent and thickness of the Pawnee coal in the study area, southeastern Montana.

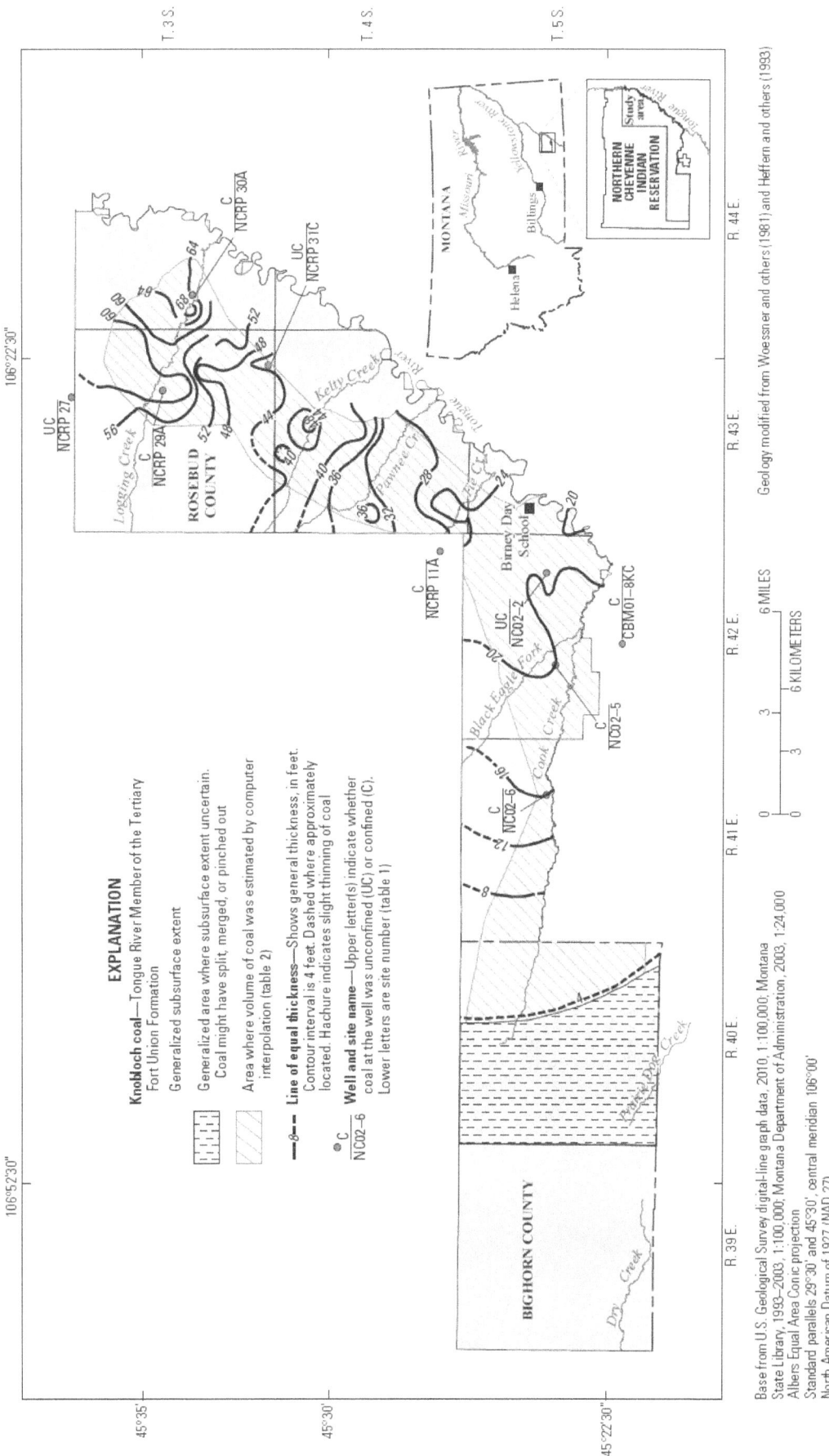

Figure 9. Generalized subsurface extent and thickness of the Knobloch coal in the study area, southeastern Montana.

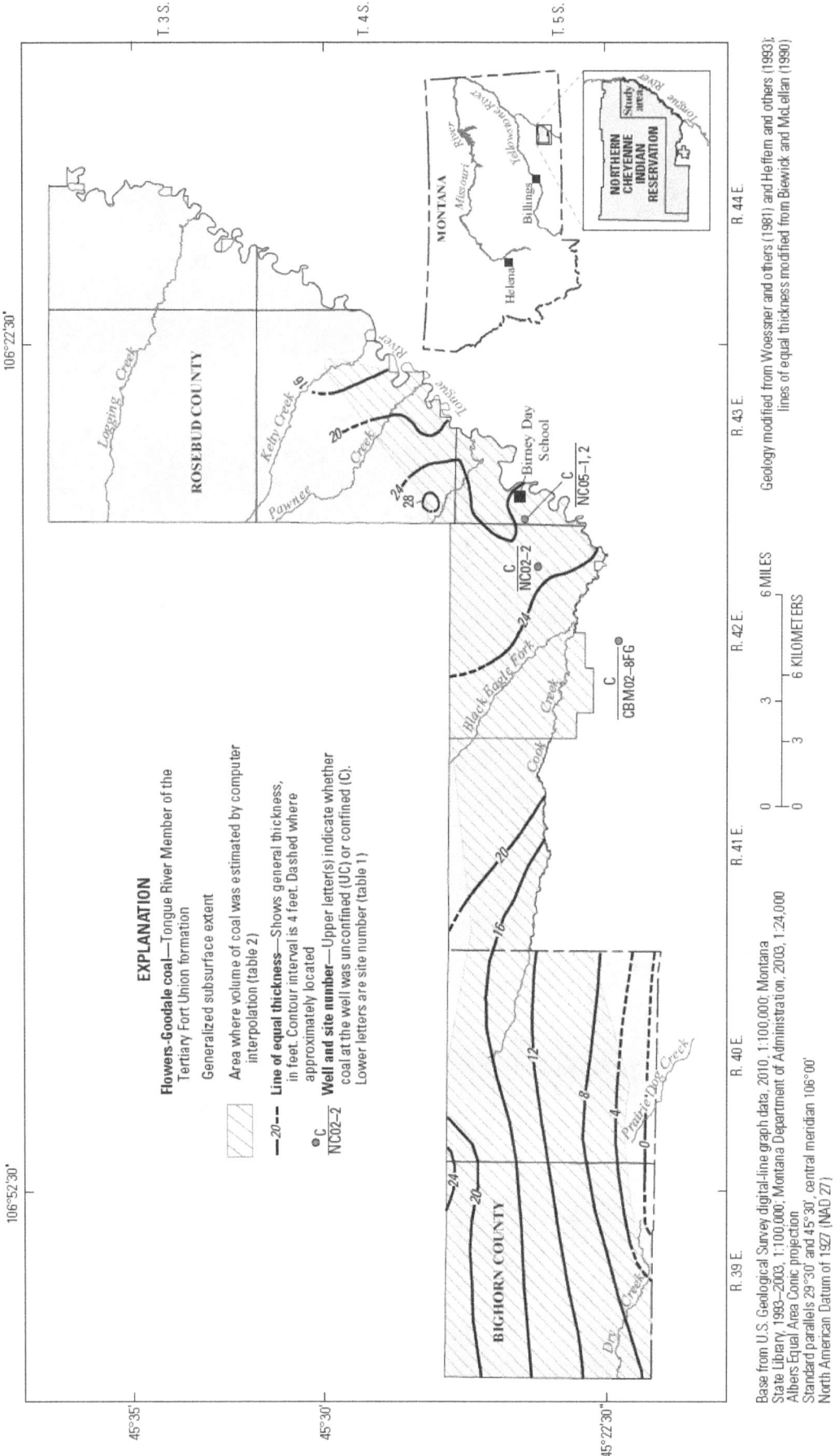

Figure 10. Generalized subsurface extent and thickness of the Flowers-Goodale coal in the study area, southeastern Montana.

Summary

Water is one of the Northern Cheyenne Tribe's most valuable natural resources—vital to the health and economic welfare of the Northern Cheyenne people. The Tongue River Member of the Tertiary Fort Union Formation is the primary source of groundwater in the Northern Cheyenne Indian Reservation, in southeastern Montana. Coal beds within this formation generally contain the most laterally extensive aquifers; in much of the reservation, a practical alternative to groundwater from the Tongue River Member does not exist.

Potential reserves of CBM might exist in the southern and eastern parts of the Northern Cheyenne Indian Reservation and in adjacent areas where some of the same coal aquifers in the Tongue River Member continue beyond the boundaries of the reservation. The extraction and subsequent management of CBM-produced water has raised concerns about the potential reduction of groundwater supplies caused by lowering of water levels and the potential effects of the disposal of produced water on surface water and soils. Consequently, the U.S. Geological Survey, in cooperation with the Northern Cheyenne Tribe, conducted a study to estimate the volume of in water in five coal aquifers. This report presents estimates of the volume of water in five coal aquifers in the eastern and southern parts of the Northern Cheyenne Indian Reservation. The Canyon, Wall, Pawnee, Knobloch, and Flowers-Goodale coal beds in the Tongue River Member of the Tertiary Fort Union Formation were investigated.

Estimates of the volume of water in the five coal aquifers were based on a specific yield of 0.01 and on the assumption that the five coal beds were unconfined. The volume of water was estimated by assuming that each coal bed was fully saturated (thickness of the coal bed was equal to the thickness of the coal aquifer, except for the Canyon coal). Estimates of the volume of water in the coal aquifers were determined only in areas where the volume of the coal beds could be estimated by computer interpolation. Only conservative estimates of the volume of water in the Canyon, Wall, Pawnee, Knobloch, and Flowers-Goodale coal aquifers are presented because (1) the subsurface extent of the coal beds are not well defined, (2) in some instances, well and drill-hole data were widely spaced and not well distributed, (3) of the possibility that coal beds split, merge, or pinch out laterally, and (4) water-level data for the five aquifers were scarce.

The Canyon coal encompasses about 40,100 acres (about 63 mi^2). Where the volume of this coal could be estimated by computer interpolation, the Canyon coal averages about 26 ft thick. The volume of water in the Canyon coal was estimated to range from about 10,400 acre-ft (75 percent saturated) to 3,450 acre-ft (25 percent saturated). Because of its high topographic position and because water-level data are unavailable to assess the amount of saturation of this coal,

100 percent saturation was not considered probable within the study area. The smaller estimates of water in the Canyon coal (75 to 25 percent saturation) are considered more reasonable. However, estimates of the volume of water in the Canyon coal might have large errors and need to be used with caution because the water-level data needed to define the volume of water were unavailable.

The Wall coal encompasses about 54,800 acres (about 86 mi^2). Where the volume of this coal could be estimated by computer interpolation, the Wall coal averages about 44 ft thick. The volume of water in the Wall coal was estimated to range from about 14,200 acre-ft (100 percent saturated) to 3,560 acre-ft (25 percent saturated). Water-level data indicate that the Wall coal was both unconfined and confined within and near the study area. Thus, the estimates of the volume of water in the Wall coal probably are reasonable.

The Pawnee coal encompasses about 73,600 acres (about 115 mi^2). Where the volume of this coal could be estimated by computer interpolation, the Pawnee coal averages about 12 ft thick. The volume of water in the Pawnee coal was estimated to range from about 9,440 acre-ft (100 percent saturated) to 2,360 acre-ft (25 percent saturated). Water-level data from one well and data from information from one geologist's log indicated that the Pawnee coal probably was fully saturated and might have been under confined conditions in the study area. However, estimates of the volume of water in the Pawnee coal might have large errors and need to be used with caution because the water-level data needed to define the volume of water were largely unavailable.

The Knobloch coal encompasses about 75,200 acres (about 117 mi^2). Where the volume of this coal could be estimated by computer interpolation, the Knobloch coal averages about 29 ft thick. The volume of water in the Knobloch coal was estimated to range from about 38,700 acre-ft (100 percent saturated) to 9,680 acre-ft (25 percent saturated). Water-level data indicate that the Knobloch coal was both unconfined and confined within and near the study area. Thus, the estimates of the volume of water in the Knobloch coal probably are reasonable.

The Flowers-Goodale coal encompasses about 132,400 acres (about 207 mi^2). Where the volume of this coal could be estimated by computer interpolation, the Flowers-Goodale coal averages about 17 ft. The volume of water in the Flowers-Goodale coal was estimated to be about 35,800 acre-ft (100 percent saturated). Water-level data from three wells and information from one geologist's log indicate that the Flowers-Goodale coal was confined at these wells. Also, because this coal generally is deeply buried in the study area, the Flowers-Goodale coal was assumed to be confined throughout its extent in the study area. Thus, the estimates of the volume of water in the Flowers-Goodale coal probably are reasonable.

Sufficient data are needed to accurately characterize coal-bed horizontal and vertical variability, which is highly complex both locally and regionally. Splitting, merging, pinching out, differential compaction, and fault displacement can further complicate correlation of coal beds and also can affect the reliability of the estimates of the volume of the coal beds between data points. Where data points are widely spaced, the reliability of estimates of the volume of coal beds is decreased. Likewise, reliable estimates of the volume of water in coal aquifers depend heavily on reliable geologic information, such as subsurface extent and thickness and correlation of coal beds. Additionally, reliable estimates of the volume of water in coal aquifers depend heavily on data from wells, such as up-to-date water-level and well-completion data and data about coal-aquifer characteristics. Because the data needed to define the volume of water were sparse, only conservative estimates of the volume of water in the five coal aquifers are presented in this report. These estimates need to be used with caution and mindfulness of the uncertainty associated with these estimates.

References Cited

Andrews, C.B., Osborne, T.J., and Whiteman, Jason, 1981, Hydrologic impacts from potential coal strip mining—Northern Cheyenne Reservation, v. II: U.S. Environmental Protection Agency Report EPA 600/ 7–80–004b, 568 p.

Baker, A.A., 1929, The northward extension of the Sheridan coal field, Big Horn and Rosebud Counties, Montana, in Contributions to economic geology, short papers and preliminary reports, part II—Mineral fuels: U.S. Geological Survey Bulletin 806–B, p. 15–67.

Bass, N.W., 1932, The Ashland coal field, Rosebud, Powder River, and Custer Counties, Montana, in Contributions to economic geology, short papers and preliminary reports, part II—Mineral fuels: U.S. Geological Survey Bulletin 831–B, p. 19–105.

Biewick, L.R.H., and McLellan, M.W., 1990, Isopach maps, perspective projections, and correlation diagrams of the Paleocene Flowers-Goodale coal resource unit in the northern Powder River Basin, Birney and Broadus 30×60 minute quadrangles, Montana-Wyoming: U.S. Geological Survey Coal Investigations Map C–136–A, scale 100,000.

Cannon, M.R., 1982, Potential effects of surface coal mining of the hydrology of the Cook Creek area, Ashland coal field, southeastern Montana: U.S. Geological Survey Open-File Report 82–681, 30 p.

Culbertson, W.C., 1987, Diagrams showing proposed correlation and nomenclature of Eocene and Paleocene coal beds in the Birney 30×60 minute quadrangle, Big Horn, Rosebud, and Powder River Counties, Montana: U.S. Geological Survey Coal Investigations Map C–113, 2 sheets.

Culbertson, W.C., and Saperstone, H.I., 1987a, Structure, coal thickness, and overburden thickness of the Wall coal resource unit, west half of the Birney 30×60 minute quadrangle, Big Horn and Rosebud Counties, Montana: U.S. Geological Survey Coal Investigations Map C–111, scale 1:100,000.

Culbertson, W.C., and Saperstone, H.I., 1987b, Structure, coal thickness, and overburden thickness of the Knobloch coal resource unit, Birney area, Big Horn, Rosebud, and Powder River Counties, Montana: U.S. Geological Survey Coal Investigations Map C–112, scale 1:100,000.

Davis, R.E., 1984, Geochemistry and geohydrology of the West Decker and Big Sky coal-mining areas, southeastern Montana: U.S. Geological Survey Water-Resources Investigations Report 83–4225, 109 p.

Derkey, P.D., 1986, Coal stratigraphy of the Lame Deer 30×60 minute quadrangle, southeastern Montana: Montana Bureau of Mines and Geology Geologic Map 43, scale 1:100,000.

Environmental Systems Research Institute (ESRI), 2010, ArcGIS, ver. 9.3.1: Redlands, Calif., Environmental Systems Research Institute, Inc., accessed April 14, 2011, at *http://www.esri.com.*

Flores, R.M., 1986, Styles of coal deposition in Tertiary alluvial deposits, Powder River Basin, Montana and Wyoming: Geological Society of America Special Paper 210, p. 79–102.

Flores, R.M., and Ethridge, F.G., 1985, Evolution of the intermontane fluvial system of Tertiary Powder River Basin, Montana and Wyoming, in Flores, R.M., and Kaplan, S.S., eds., Cenozoic paleogeography of the west-central United States, Rocky Mountain Paleogeography Symposium 3, Rocky Mountain Section: Denver, Colo., Society of Economic Paleontologists and Mineralogists, p. 107–126.

Flores, R.M., Spear, B.D., Kinney, S.A., Purchase, P.A., Gallagher, C.M, 2010, After a century—Revised Paleogene coal stratigraphy, correlation, and deposition, Powder River Basin, Wyoming and Montana: U.S. Geological Survey Professional Paper 1777, 97 p., CD–ROM in pocket.

Freeze, R.A., and Cherry, J.A., 1979, Groundwater: Englewood Cliffs, N.J., Prentice-Hall, 604 p.

Groenewold, G.H., Hemish, L.A., Cherry, J.A., Rehm, B.W., Meyer, G.N., and Winczewski, L.M., 1979, Geology and geohydrology of the Knife River Basin and adjacent areas of west-central North Dakota: North Dakota Geological Survey Report of Investigation 64, 402 p.

Gruber, J.R., Jr., 1990, Coal geology and resources of the Kirby quadrangle, Big Horn County, Montana: Montana Bureau of Mines and Geology Geologic Map 52, 30 p, scale 1:24,000.

Hansen, W.B., and Culbertson, W.C., 1985, Correlated lithologic logs and analyses of 1982 coal drilling in Big Horn, Prairie, Rosebud, and Treasure Counties, Montana: U.S. Geological Survey Open-File Report 85–738, 125 p.

Heath, R.C., 1983, Basic ground-water hydrology: U.S. Geological Survey Water-Supply Paper, 84 p.

Heffern, E.L., 1980, Coal stratigraphy of the Tongue River Member, Northern Cheyenne Reservation, Montana, *in* Carter, L.M., ed., Geology of Rocky Mountain coal, Proceedings of the 4th Symposium, Resource Series 10: Denver Colo., Colorado Geological Survey Department of Natural Resources, p. 76–80.

Heffern, E.L., and Coates, D.A., 2004, Geologic history of natural coal-bed fires, Powder River Basin, USA: International Journal of Coal Geology, v. 59, p. 25–47 (also available at *http://dx.doi.org/10.1016/j.coal.2003.07.002*).

Heffern, E.L., Coates, D.A., Whiteman, Jason, and Ellis, M.S., 1993, Geologic map showing distribution of clinker in the Tertiary Fort Union and Wasatch Formations, northern Powder River Basin, Montana: U.S. Geological Survey Coal Investigations Map C–142, scale 1:175,000.

Heffern, E.L., Reiners, P.W., Naeser, C.W., and Coates, D.A., 2007, Geochronology of clinker and implications for evolution of the Powder River Basin landscape, Wyoming and Montana, *in* Stacher, G.B., ed., Geology of coal fires— Case studies from around the world: Geological Society of America Reviews in Engineering Geology, v. 16, p. 155–175 [also available at *http://dx.doi.org/101130/2007.4118(10)*].

Hopkins, W.B., 1973, Water resources of the Northern Cheyenne Indian Reservation and adjacent area, southeastern Montana: U.S. Geological Survey Hydrologic Investigations Atlas HA–468, scale 125,000.

Kennelly, P.J., and Donato, Teresa, 2001, Hydrologic features of the potential coalbed methane development area of the Powder River Basin, Montana: Montana Bureau of Mines and Geology Open-File Report MBMG 448, scale 1:200,000.

Lewis, B.D., and Roberts, R.S., 1978, Geology and water-yielding characteristics of rocks of the northern Powder River Basin, southeastern Montana: U.S. Geological Survey Miscellaneous Investigations Series Map I–847–D, scale 1:250,000.

Lohman, S.W., 1979, Ground-water hydraulics: U.S. Geological Survey Professional Paper 708, 70 p.

Mapel, W.J., 1976, Geologic map and coal sections of the Birney quadrangle, Rosebud County, Montana: U.S. Geological Survey Miscellaneous Field Studies Map MF–813, scale 1:24,000.

Matson, R.E., and Blumer, J.W., 1973, Quality and reserves of strippable coal, selected deposits, southeastern Montana: Montana Bureau of Mines and Geology Bulletin 91, 135 p.

McClymonds, N.E., 1982, Hydrology of the Prairie Dog Creek drainage basin, Rosebud and Big Horn Counties, Montana: U.S. Geological Survey Water-Resources Investigations Report 81–37, 64 p.

McLellan, M.W., Biewick, L.R.H., Molina, C.L., and Pierce, F.W., 1990, Cross section showing the reconstructed stratigraphic framework of Paleocene rocks and coal beds in the northern and central Powder River Basin, Montana and Wyoming: U.S. Geological Survey Map I–1959–A, scale 1:500,000.

McKay, E.J., 1976a, Preliminary geologic map and coal section of the Willow Creek quadrangle, Rosebud and Powder River Counties, Montana: U.S. Geological Field Studies Map MF–802, scale 1:24,000.

McKay, E.J., 1976b, Preliminary geologic map and coal section of the King Mountain quadrangle, Rosebud and Powder River Counties, Montana: U.S. Geological Field Studies Map MF–817, scale 1:24,000.

Meredith, Elizabeth, Kuzara, Shawn, Wheaton, J.R., Bierbach, Simon, Chandler, Kevin, Donato, Teresa, Gunderson, Jay, and Schwartz, Charles, 2010 Annual coalbed methane regional groundwater monitoring report—Powder River Basin, Montana: Montana Bureau of Mines and Geology Open-File Report MBMG 600, 107 p.

Microsoft Corporation, 2010, Microsoft Office software: Redmond, Wash., Microsoft Corporation, accessible at *http://www. microsoft.com.*

Morin, R.H., 2005, Hydrologic properties of coal beds in the Powder River Basin, Montana, I— Geophysical log analysis: Journal of Hydrology, v. 308, p. 227–241 (also available at *http://doi:10.1016/j.jhydrol.2004.11.006*).

Montana Department of Natural Resources and Conservation, 1976, River mile index of the Yellowstone River: Helena, Mont., Montana Department of Natural Resources and Conservation, Water Resources Division, 61 p.

Montana Department of Natural Resources and Conservation, 2002, Northern Cheyenne Settlement Agreement: Montana Department of Natural Resources and Conservation, Trust Land Management Division, accessed June 20, 2013, at *http://dnrc.mt.gov/Trust/MMB/ OtterCreek/2General/NorthernCheyenneSettlement/ NCT%20Settlement%20Agreement.pdf*

Purl, R., Evanoff, J.C., and Brugler, M.L., 1991, Measure-ment of coal cleat porosity and relative permeability characteristics [abs.]: Society of Petroleum Engineers Gas Technology Symposium, Houston, Tex., January 22–24, 1991, Society of Petroleum Engineers, accessed July 28, 2010, at *http://www.onepetro.org/mslib/servlet/ onepetropreview?id=00021491&soc=SPE.*

Rehm, B.W., Groenewold, G.H., and Morin, K.A., 1980, Hydraulic properties of coal and related materials, Northern Great Plains: Ground Water, v. 18, no. 6, p. 551–561.

Renick, B.C., 1929, Geology and ground-water resources of central and southern Rosebud County, Montana, *with* Chemical analyses by H.G. Riffenberg: U.S. Geological Survey Water-Supply Paper 600, 140 p.

Robinson, L.N., and Culbertson, W.C., 1984, Map showing isopachs of coal and overburden of the Canyon coal bed, Birney 30×60 minute quadrangle, Big Horn, Rosebud, and Powder River Counties, Montana: U.S. Geological Survey Coal Investigations Map C–96–B, scale 1:100,000.

Slagle, S.E., Lewis, B.D., and Lee, R.W., 1985, Ground-water resources and potential hydrologic effects of surface coal mining in the northern Powder River Basin, southeastern Montana: U.S. Geological Survey Water-Supply Paper 2239, 34 p.

Stoner, J.D., 1981, Horizontal anisotropy determined by pumping in two Powder River Basin coal aquifers, Montana: Ground Water, v. 19, no. 1, p. 34–40.

U.S. Geological Survey, 2010a, Water resources data for the United States water year 2010: U.S. Geological Survey Water-Resources Data Report WDR-US-2010, accessed August 3, 2011, at http://wdr.water.usgs.gov/wy2010/search.jsp.

U.S. Geological Survey, 2010b, Water-resources data for the United States water year 2010: U.S. Geological Survey Water-Resources Data Report WDR-US-2010, accessed April 4, 2011, at http://wdr.water.usgs.gov/wy2010/pdfs/452411106301601.2010.pdf.

U.S. Geological Survey and Montana Bureau of Mines and Geology, 1980, Coal drilling during 1978 in Big Horn, Daniels, Dawson, Musselshell, Richland, Roosevelt, Rosebud, Valley, and Yellowstone Counties: U.S. Geological Survey Open-File Report 80–267, 217 p.

Van Voast, W.A., and Hedges, R.B., 1975, Hydrogeologic aspects of existing and proposed strip coal mines near Decker, southeastern Montana: Montana Bureau of Mines and Geology Bulletin 97, 31 p.

Vuke, S.M., Heffern, E.L., Bergantino, R.N., and Colton, R.B., 2001a, Geologic map of the Lame Deer 30×60 minute quadrangle, eastern Montana: Montana Bureau of Mines and Geology Open-File Report MBMG 428, scale 1:100,000.

Vuke, S.M., Heffern, E.L., Bergantino, R.N., and Colton, R.B., 2001b, Geologic map of the Birney 30×60 minute quadrangle, eastern Montana: Montana Bureau of Mines and Geology Open File Report 431, scale 1:100,000.

Warren, W.C., 1960, Reconnaissance geology of the Birney-Broadus coal field, Rosebud and Powder River Counties, Montana: U.S. Geological Survey Bulletin 1072–J, p. 561–585.

Weeks, E.P., 2005, Hydrologic properties of coal-beds in the Powder River Basin, Montana, II—Aquifer test analysis: Journal of Hydrology, v. 308, p. 242–257 (also available at http://dx.doi.org/10.1016/j.hydrolo.2004.11.002).

Western Regional Climate Center, 2010, Western U.S. climate summaries [eastern Montana], accessed September 20, 2010, at http://www.wrcc.dri.edu/summary/Climsmemt.html.

Wheaton, J.R., and Donato, T.A., 2004, Ground-water monitoring program in prospective coalbed-methane areas of southeastern Montana—Year One: Montana Bureau of Mines and Geology Open-File Report MBMG 508 [variously paged].

Wheaton, J.R., and Metesh, John, 2002, Potential ground-water drawdown and recovery from coalbed methane development in the Powder River Basin, Montana: Montana Bureau of Mines and Geology Open-File Report MBMG 458, 59 p.

Wo, Shaochange, Lopez, D.L., and Whiteman, Jason, Sr., 2004, Northern Cheyenne Reservation coal bed natural resource assessment and analysis of produced water disposal options: Idaho Falls, Idaho, Idaho National Engineering and Environmental Laboratory Report INEEL/EXT–04–02201, 45 p.

Woessner, W.W., Osborne, T.J., Heffern, E.L., Andrews, Charles, Whiteman, Jason, Spotted Elk, Wesley, and Morales-Brink, Daniel, 1981, Hydrologic impacts from potential coal strip mining—Northern Cheyenne Reservation, v. I: U.S. Environmental Protection Agency Report EPA 600/7-81-004a, 302 p.

Wood, G.H., Kehn, T.M., Carter, D.M., and Culbertson, W.C., 1983, Coal resource classification system of the U.S. Geological Survey: U.S. Geological Survey Circular 891, 65 p.

Woods, A.J., Omernik, J.M., Nesser, J.A., Shelden, J., Comstock, J.A., and Azevedo, S.H., 2002, Ecoregions of Montana (2d ed.): U.S. Environmental Protection Agency Western Ecology Division, accessed October 8, 2009 at http://www.epa.gov/naaujydh/pages/ecoregions/mt_eco.htm.

Publishing support provided by:
Denver Publishing Service Center, Denver, Colorado

For more information concerning this publication, contact:
Director, Montana Water Science Center
U.S. Geological Survey
3162 Bozeman Avenue
Helena, MT 59601
(406) 457-5900

Or visit the Montana Water Science Center Web site at:
http://mt.water.usgs.gov/

This report is available at: http://pubs.usgs.gov/sir/2012/5209/

Tuck and others—Estimates, Water in Five Coal Aquifers, Northern Cheyenne Indian Reservation, Southeastern Montana—Scientific Investigations Report 2012–5209

www.ingramcontent.com/pod-product-compliance
Lightning Source LLC
Chambersburg PA
CBHW081411170526
45166CB00010B/3298